GOATS

GOATS

A GUIDE TO MANAGEMENT

PATRICIA ROSS

The Crowood Press

First published in 1989 by
The Crowood Press Ltd
Ramsbury, Marlborough
Wiltshire SN8 2HR

Paperback edition 1995

British Library Cataloguing in Publication Data

A catalogue record for this book is available from the British Library.

ISBN 1 85223 912 3

Acknowledgements
All photographs by Anna Oakford, except those on pages 12 (top), 13, 14, 16, 18, 21, 22, 84, 86 and 91, which are by Patricia Ross.

Line-drawings by Claire Upsdale-Jones.

Typeset by PCS Typesetting, Frome, Somerset
Printed in Great Britain by Redwood Books, Trowbridge, Wiltshire

Contents

Introduction

Man has been keeping goats for some 10,000 years – ever since our ancestors decided that it was more sensible to keep animals tethered than to waste time and energy hunting them. The numerous varieties of the modern domestic goat are all descended from *Capra aegagrus*, a species common throughout the East, and also, in some cases, from crosses with *Capra falconeri* of the Himalayas, or *Capra prisca* from the Mediterranean basin.

Throughout history, the goat has been valued for its milk and its meat, although sadly since kid meat is akin to veal in quality – goat meat is now looked down on in the 'sophisticated' Western countries. In more enlightened lands, the goat is still a highly prized domestic animal and is found in greatest concentrations in India, parts of Africa and China. Surprising though it may seem, more people round the world drink goat's milk than cow's milk.

The reasons for the goat's popularity are easy to understand. To begin with, goats have much tougher mouths than their two main ruminant rivals, the cow and the sheep. This means that they can consume a very wide range of plant material and thrive where the other two would have difficulty surviving. Also, pound for pound in body weight, a good goat can produce twice as much milk as a cow. Taken together, these facts mean that a goat costs less to keep and produces more than its traditional rival, the cow. Since they are so much smaller than cows, goats are much less expensive to buy. Furthermore, as they have a much longer lactation period, two or three goats can provide a constant supply of milk when, with one cow, you are faced with the problem of having either a superabundant feast of milk or two months of famine.

Another fact which has been given more prominence in recent years is that goat dairy products are less likely to trigger off allergic reactions in humans than those from the cow.

The uses of a goat are not, of course, restricted to providing meat and milk. The milk itself can form the basis for a range of products, from yoghurt, butter and cream to delectable cheeses. I have even found that goat's cream can be used as the base for a very effective skin cream! Some breeds, such as the Angora, are kept for their hair. The Angora's mohair is used for spinning wool and weaving fabrics, and the

Introduction

skin, complete with fleece makes a handsome rug or can be fashioned, like sheepskin, into coats. Few people contemplating goat-keeping will anticipate breeding goats for their skins, but these provide some of the finest leather – glacé kid shoe leather, or the morocco leather used in bookbinding and high-quality leatherwork. Several owners not only keep their goats for produce, but also derive a great deal of satisfaction from showing their animals, and I shall go into this aspect of goat-keeping in Chapter 8.

It may seem from the above that there can be no drawbacks to keeping a goat. Unfortunately, this is not the case. One of the main prerequisites for owning goats is time. You may find that they become favourite family pets, but they should not be viewed as such. A doe in lactation will need milking twice a day, every day of the week, every day of the year. It is no good ignoring the fact. Who will look after them when you go on holiday? Similarly, you cannot leave a goat in your back garden and expect it to look after itself and *not* eat the roses or prize young trees – it will prefer them to your overgrown lawn, I'm afraid. Staking the animal is no solution, as your goat might easily strangle itself. If you want to restrict its diet – and its access to surrounding gardens or the open country – there is really no alternative to penning.

8

Grazing goats contained by stock-proof wire.

Then there is the smell. Contrary to popular belief, not all goats smell. Does (females) are inoffensive, but bucks (males) give off what is to many people a distinctly unpleasant odour which is at its most pungent during the autumn/winter breeding season. If you are planning to keep goats within smelling distance of neighbours you can solve the problem by keeping only does. At this point I should also mention that goats have a herding instinct, so keeping a solitary goat is not the kindest thing to do. A pair will be much happier.

Having pointed out the drawbacks, I am sure that you will decide that these are far outweighed by the advantages. Certainly, my goats have been an invaluable source of satisfaction – and profit – over the years.

1 Breeds

Over the centuries, many 'authorities' tried to classify the enormous variety of goats throughout the world into a rationalised system of breeds. The criteria they used for identifying breeds included hair type, length and shape of horns, shape of ears and even, in one gentleman's opinion, eye colour. Whatever the criteria used, the classification was open to dispute. Fortunately, the creation of the British Goat Society and the evolution of its Herd Books have provided amateur and professional goat-breeder alike with a common reference point.

ANGLO-NUBIAN

The Anglo-Nubian descends from goats imported into England from the East in about 1850, which were subsequently crossed with various native and other imported goats. The breed was first recognised in the 1890s and has fluctuated in popularity, although recent improvements in the breed type and its milk yield have brought it back into favour. Another factor in this renewal of interest has been an increase in export demand.

An Anglo-Nubian goat is a tall, heavy animal – the female can weigh up to 200lb (90kg) and the male up to 300lb (135kg). The most distinctive features are its short, fine coat and long, pendulous, low-set ears. Other characteristics include the Roman nose, straight and high back, and the similarly set tail. Anglo-Nubians can come in any colour or combination of colours, but facial streaks indicate Swiss blood and are frowned upon.

One disadvantage with this breed is that it usually has a much shorter lactation period than other breeds, although the increased butterfat and protein in an Anglo-Nubian's milk is a compensatory factor.

ANGORA

It seems that the Angora (or Mohair) goat was originally imported into Asia Minor from the mountains of Tibet, and the first pair to arrive in Europe reached the Dutch court as early as 1541.

Anglo-Nubian. Note the long, low-set ears and Roman nose.

Angora doe.

Obviously, Angora goats are kept principally for their luxurious white fleece rather than for their milk (which is nevertheless very nutritious). However, it has been found that the quality of this fleece deteriorates where the rainfall exceeds 20in (50cm) a year. This disadvantage is offset to some extent by the fact that the meat of the Angora is superior to that of any other breed.

Apart from the distinctive fleece, the main characteristics of the Angora are a fine head, with the fleece growing well over the forehead, wide, thin ears and flat-shaped horns.

BRITISH ALPINE

The geographical sources of this breed's ancestors are open to question, but the first notable herd of these goats had been established by 1911, and within a decade the British Goat Society was referring to the breed as the British Alpine.

This short-haired goat is sometimes thought of as a 'Black Toggenburg'. Certainly it has a very striking appearance, with its dramatic jet-black colouring and white Swiss markings (facial stripes and markings on the ears, legs, rump and below the tail). It is tall and

British Alpine. Note the Swiss markings.

rangy, with a long, lean head, straight face, long slender neck and erect ears that point slightly forward. The British Alpine has a fair milk yield of some 10lb (4.5kg) a day.

GOLDEN GUERNSEY

As well as having a common geographical origin, the Guernsey goat and the Guernsey cow share another distinctive feature – their colour. It is only relatively recently that the breed was officially recognised on its home territory (the Guernsey Goat Society started its own Herd Book for the breed in 1922) and the Golden Guernsey has had to fight for survival. Fortunately, a trust fund was established in the Channel Isles and the work of some dedicated breeders has made this goat's future look considerably healthier.

A Golden Guernsey has not only golden hair, but also golden skin. The goat is small and compact, weighing anything from 120–200lb (55–90kg), depending on sex.

Golden Guernsey.

ENGLISH GUERNSEY

A Golden Guernsey female may be mated to a Saanen or British Saanen male. It will then take several years of crossing subsequent generations to the Golden Guernsey or English Guernsey male to produce the upgraded English Guernsey. These goats may be paler in colour and larger than the Golden Guernsey, but hopefully they will have a better milk yield.

OLD ENGLISH

The origin of the Old English goat dates back as far as 7000 BC, or even further, to the time when Neolithic man in Asia Minor changed from being a hunter and gatherer into a more settled farmer. Part of this process was the domesticating of wild animals – mostly sheep and goats – to provide a constant source of meat. Descendants of these goats were eventually imported into Britain where they reigned supreme for centuries, until the arrival of more productive breeds. The Old English was then relegated to the bottom of the pile, except for cross-breeding purposes. Its decline was so swift that by 1927 very few were left and, to all intents and purposes, the Old English breed was considered extinct.

However, thanks to some timely and vigorous conservation work, some of these stocky, rugged little goats survive, both in the wild and in private herds. The Irish, Scottish and Welsh goats suffered similar fates, although the Welsh species is now truly extinct. The Irish can be found in feral herds in all four countries, and Scotland has a healthy population of feral goats – including some pure white herds – scattered over the country.

The typical modern-day Old English goat is small, with long swept back horns, a shaggy coat and short legs. The hair varies in colour but is commonly a fawn or blue roan. Although the Old English does not produce large quantities of milk, it is a robust animal.

SAANEN

This breed takes its name from the Saane valley in the Swiss canton of Berne, although it is also found in considerable numbers in the nearby Simmental valley. The first examples were imported into Britain in the early 1900s, and the Saanen has since gone from strength to strength. There were a few breeding problems along the way, with an unaccept-

Saanen.

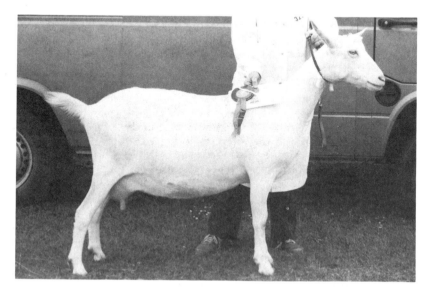

British Saanen.

able level of infertility and hermaphroditism among hornless Saanen kids, but this was resolved by the acceptance of the registration of horned males.

The Saanen is of average size – the adult female weighs about 150lb (70kg) – and is white with a short coat which often has a long fringe along the back and down the hindquarters. Other notable features are a long, deep body, a slender neck and erect ears which point slightly forward. The long lactation period, generous milk yield and placid nature of the Saanen make it a popular dairy goat.

BRITISH SAANEN

As its name implies, the British Saanen breed was the result of crossing the pure Saanen with other breeds. By 1925 – three years after the first Saanens were introduced into the United Kingdom – sufficient breeding progress had been made for the British Saanen to be allocated a section in the Herd Book, and by 1943 the breed was firmly established.

The British Saanen is much more popular than its pure ancestor, chiefly because it has the longest lactation and highest milk yield of all the dairy goats. Like the pure Saanen, the British breed is white, has erect, forward-pointing ears and retains the attractive placidity which makes it ideal for free range or stall-feeding. On the other hand, it is longer in the leg and larger, with a doe weighing approximately 170lb (77kg).

TOGGENBURG

The 'Tog', as it is often called, is an endearing little goat. The breed originated in Switzerland, with the first examples reaching England in the mid–1880s. Since then Toggenburgs have been cross-bred to produce the British Toggenburg which is officially recognised as a separate breed. Only the imported goats or their pure-bred progeny can be registered as Toggenburgs.

The docile, sweet-natured Toggenburg is compact and attractive with short-to-medium length hair. The colour of this hair ranges from fawn to dark brown, and there are white facial stripes and white markings on the ears, legs, rump and under the tail. Although not one of the most prolific milk-producers, the Toggenburg makes up for this with a robust constitution and undemanding nature.

Toggenburg.

British Toggenburg. Her coat is slightly darker and smoother than her pure namesake.

BRITISH TOGGENBURG

It was in 1925 that this breed was officially classified, after consistent cross-breeding of pure Toggenburg males with Swiss and English females had produced a larger, more productive milk goat which was proving satisfactorily true to type.

The British Toggenburg shares its pure namesake's colouring and white Swiss markings, but is sometimes a darker brown. The hair is slightly shorter and other distinguishing features are a larger frame, longer ears and a straighter nose. The reputation for high milk yields is well deserved and this, combined with the robustness of the British Toggenburg, make it a good choice for the novice.

BRITISH

The British goat is a cross between two of the pedigree breeds, for example, British Saanen and British Toggenburg.

2 Selecting a Goat

Before you rush off in a wave of enthusiasm to buy your first goat, it would be as well to ensure that you are fully equipped to care for your purchase (*see* Chapter 3) and to arm yourself with some basic information. A good starting point is to contact the British Goat Society who will give you the address of your local affiliated goat club. These clubs are a mine of information and you will be able to seek the advice of experienced goat-keepers.

By now you will probably have formed a preference for a certain breed. Physical attributes and average milk yields are not necessarily the only criteria to use in assessing a breed. If this is your first experience of goat-keeping, you will also need to consider temperament. The Saanen types and the Toggenburg are generally quite placid, as is the Golden Guernsey. The Saanen type has an extra advantage in that it is perfectly happy being restricted and is therefore an ideal garden

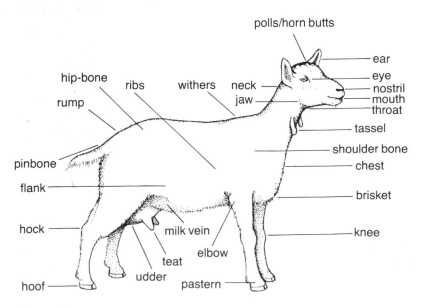

The goat's anatomy.

goat – although it will still need to have some exercise. The British Toggenburg can be boisterous. Alpines, despite their friendly nature, are less tractable, need a firmer hand and have an alarming ability to surmount fences. As for Anglo-Nubians, they can be excessively noisy with their strange bleat and aggressive growl when annoyed.

FEATURES TO LOOK FOR

If you want a good milk goat, the points to look for are common to all breeds. She will have a long, neat head, tapering towards the muzzle from a broad forehead, and her horns, if she has any, should be small, thin and uniform. She should look alert, have large, bright eyes and a thoroughly feminine face.

The other areas to which special attention should be paid are the body, udder and teats. As far as the body is concerned, it should be long, and she should have a large, deep frame and well-rounded ribs, allowing a good stomach capacity. A good milker is always wedge-shaped – much deeper at the hindquarters than at the chest. The back

British Saanen showing all the signs of a well-bred goat.

21

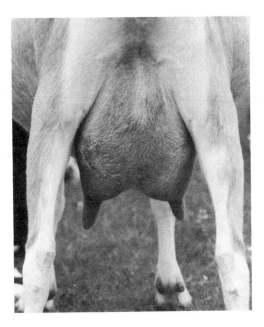

A good udder with the teats set nicely apart and pointing forwards slightly.

should be straight, with a gentle slope from the hips to the rump and wide-set, straight hocks. Do not assume that a thin goat is to be avoided. On the contrary – a fat goat is probably putting all its energy into meat production, whereas a thin goat may be converting the bulk of her food into milk.

The udder should be soft, round and well-attached, with no lumps or skin disorders. When quite full it will stretch to a considerable size, but should shrink back after milking. The teats should be tapering, set nicely apart and point forwards slightly. It is a good idea to attend one or two milkings to check on her yield and to see whether she is easy to handle. It is also advisable to taste the milk for flavour.

Soft, supple skin and fine, soft hair are signs that the doe is in good condition. Look for sound teeth and good neatly-shaped feet that have been trimmed regularly as evidence that the animal has been well cared for.

FEATURES TO AVOID

First and foremost, do not let either your sentimentality or any bargain-hunting instinct get the better of you. Sometimes the culls, rather than the best of a herd, are put up for sale and, although you may feel that all they need is some fattening up and a dose of tender loving care, they will never be as productive or healthy as prime livestock, and any weakness may be passed down to their progeny.

Other features which should be avoided include noticeably white skin under the tail and the eyelids, as these are signs of anaemia which could be caused by a severe infestation of internal parasites. General pointers to an unhealthy animal are lethargy, a poor coat, dull eyes, and foul breath.

A broad chest may indicate a tendency to accumulate meat and fat rather than milk, although it might simply be indicative of a good constitution – check the milk yields. A masculine, wide face is often the sign of a poor milker. A steep slope to the tail, a raised back, narrowness between the hips and lack of depth in the body are all undesirable, as are poorly formed legs and misshapen udders and teats.

AGE

Opinions differ on what age of goat to aim for. Some believe that the best age is three years, just after it has borne its second litter, since the milk yield after the first litter is relatively small. Others feel that it is preferable for the novice to purchase a younger animal, and even a goatling (a one-year-old goat) in kid, so that there is time to become used to the animal and to caring for it before the kid – and the milk – arrives. This does have the advantage of providing the mother-goat with a companion, but also has the disadvantage that, when the time comes to milk her, it will be a novel experience for both the owner *and* the goat – not an ideal combination! There is a strong argument for buying a mature milker who has recently kidded and a goatling who is ready to breed, as they will be companions for each other and you will also have a constant supply of milk if you breed them in alternate years.

Whatever your choice, unless the goat is registered (*see* below) you will need to know how to check its age. A one-year-old will have the full complement of thirty-two teeth (eight incisors and twelve molars in the lower jaw, plus twelve molars in the upper jaw). In the second year the two centre incisors fall and two new, considerably larger, incisors replace them. In the third year the two incisors flanking these new teeth are replaced, followed by the pair flanking these four large incisors in the fourth year. Finally, the remaining two milk incisors are replaced in year five, so that the goat has a 'full mouth'. Thereafter, the age of the goat can be judged by the wear on the teeth, although the degree of wear will depend, to some extent, on diet.

REGISTRATION

All goats eligible for inclusion in the relevant section of the British Goat Society's Herd Book will have been duly registered with the Society by their breeders and, when you buy or view such a goat, you must ask for the registration card. This will tell you the name of the goat, its birth date, its registration number, its ear number – if it has one, in which case it should be compared with the mark in its ear – its parents and the name of the breeder. If the person who is offering the goat for sale is not the breeder, then the seller's name should be the last one on the card. If it is not, make sure that the error is corrected before agreeing to purchase the goat. The person selling the goat will send a transfer form together with the goat's Registration Card to the BGS and the altered card will be returned to the person buying the goat. (The seller will pay for the transfer.)

Is a Registered Goat Worth the Extra Money?

This is a moot point. When buying a goat, it is advisable to give some consideration to the points for and against buying a registered one. There is no doubt that a pedigree animal will cost more than a cross-breed. However, even if you are buying a goat simply for milking, a registered milker will cost the same to maintain as a 'bargain' purchase from 'a friend of a friend', and since these goats have been bred for high yields and extended lactation you will have some guarantee that she will give you a plentiful supply of milk over a long period. In contrast, although you may feel that everything points to your non-registered goat being a promising specimen, you may well be sadly disappointed.

It is also worth bearing in mind that a solid, but unremarkable specimen which is backed up by an impressive breed line is a much sounder basis from which to start breeding your own goats, than a freak marvel that was sported from unregistered stock and whose offspring are therefore liable to revert to type. At the same time, some excellent cross-bred milkers have been produced in recent years. If you are considering purchasing such a goat, I can only suggest that you obtain an independent opinion on the animal before going ahead. On balance, as with most major purchases, you would do well to aim as high as your budget will allow.

A pedigree goat is worth the extra money.

WHERE TO FIND YOUR GOAT

The members of your local goat club and the British Goat Society's monthly journal are rich sources of information on goats for sale. You may also see some interesting animals at shows. It is best to buy your goats from a reputable breeder – especially if you are inexperienced – to be sure that you are investing in a sound goat of good parentage.

3 Housing and Equipment

Before you even think about purchasing a goat, you will have to spend a considerable amount of time and energy, plus some money, in providing your livestock with shelter and in acquiring the basic equipment for goat husbandry. As with any occupation or hobby, you will derive a great deal more pleasure from the activity if you use the correct tools and have clean, well-ordered surroundings in which to work.

THE GOAT-HOUSE

It may seem strange that goats need shelter at all. After all, do they not survive quite happily in even the most inhospitable parts of Britain? Indeed they do, but only by finding their own shelter – caves, overhangs or abandoned huts – where they make dry, warm beds of accumulated droppings.

The amount of wear a goat-house gets depends to a great extent on the climate. In the tempestuous north goats may spend a large proportion of their time indoors, while in the temperate south most of their day is spent outdoors, with the goat-house serving purely as a dormitory. It follows that a southern goat-house can be constructed to lower standards than one in the north.

Which System

The use to which a goat-house is put can influence its layout and amenities. In warm, dry countries, nearly all such houses are communal, with simple sleeping benches and no penning, whereas, in wetter climates where goats spend more time inside, pens are more frequently used. The former system would seem perfectly adequate, even in Britain, and it does have significant advantages – the goats have more freedom and, as they are such social animals, are more content. However, these benefits are offset by some considerable drawbacks. One is that the goats have to be isolated when being fed concentrates

Penned goat-house showing wide gangway and milk scales.

to ensure that each goat gets its ration. Another is that it is common, particularly when animals are being bought and sold quite frequently, or when some are horned and others are hornless, for bullying to develop as the protagonists contest leadership of the flock.

Most goatkeepers in Britain opt for a penned house. Although this needs to be somewhat larger to make sure that each goat has enough living room, it does make the animals easier to handle and, providing they can communicate with, and see each other, they are not unhappy.

There are two other systems which should be mentioned in passing – hecks and stalls. With hecks, the goats are severely restricted and are held by the neck when feeding. This is a space-saving arrangement, but the goats will not like it and their yield will be severely affected. It is only really practicable if the goats are allowed plenty of freedom for the majority of the day, and this is only possible in much warmer, drier climates. Stalls are preferable to hecks, giving the goats more room in which to move, but even here they are tied. If your animals are to be kept under shelter most of the time, they will be much happier in a more spacious pen.

Structure

Unfortunately, modern-day housebuilders have no truck with the versatile outbuildings of our grandparents' day. 'Utility' rooms and elegant glass conservatories may be more decorative than the trusty scullery annexe and garden shed, but they are obviously not the ideal quarters for a goat. Be that as it may, many householders still have access to some kind of outbuilding that could, with a little thought and some hard work, be converted into a perfectly satisfactory goat-house. Garages, sheds and barns can all make suitable houses.

The most important criteria are adequate light and ventilation, and the absence of draughts and damp. A southerly aspect would be ideal, providing a little extra warmth as well as light. If the building to be used adjoins your house, the goats will benefit from another free source of heat in the depths of winter. You may think that an animal which can survive in the wild, in all weathers, would not mind the odd draught and dampness, but goats are prone to pneumonia so giving them this protection is not only considerate, but in your interests too. Similarly, insufficient ventilation makes any animal susceptible to disease.

On the subject of ventilation, it is worth mentioning that if your goats are going to spend a large amount of time in the open, there should be plenty of fresh air circulating in the house. Conversely, if they are only going to venture outside for short exercise periods, the

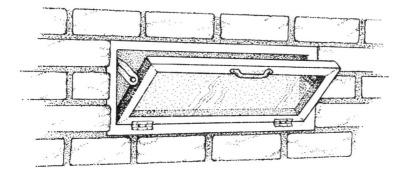

*A hopper-style window, opening inwards and at least
7ft (230cm) from the ground.*

circulation of cold air in the house should not be too bracing. There is
a similarity with horticulture – even the hardiest of plants will have
difficulty surviving if it is raised in a hothouse and then plunged
straight outside to withstand the vagaries of the British climate.

Size

An *ideal* goat-house will be spacious, with plenty of room not just for
the goats but also for you to move around in comfort. The size of the
house will depend on how many animals it is to shelter. Let us assume
that you are thinking of a fairly basic structure with pens, either converted from an old outbuilding or built from scratch with the
minimum of extras.

Each milker will need a pen measuring at least 5ft (150cm) square
with side walls no lower than 4ft (120cm) high. A 4ft (120cm) ob-
stacle should deter the most agile goat from leaping over into an
adjoining pen but if you should want to increase the height, use a
transparent barrier such as mesh or rails, so that your animals will still
be able to see each other – communication is very important to a herd
animal's well-being. It is also worth remembering that, although a partition may be 4ft (120cm) high from the floor, deep litter can reduce
the effective height quite considerably.

You will need to allow for a wide gangway in which you can comfortably manoeuvre a wheelbarrow when mucking out. Other space-
consuming requirements are a food store where you can keep hay,
straw and tools close at hand, and a sectioned-off milking parlour of

similar size to the goat pens, although you could milk your charges in the gangway, or even in your kitchen. Whatever your solution to this problem, you will need some space for a dairy, either in the goat-house or within easy reach of it.

Outside the house, you must provide an exercise yard. This can be of earth or concrete, but in either case it should slope very slightly for efficient drainage.

Materials

Base

There is some disagreement over the best base for a goat-house. Wooden floors are obviously not ideal, given the large amount of wet bedding that will sit on them, but if your outhouse already has a wooden floor, there is no point in taking it up – just leave it to rot away quietly. Many goatkeepers swear by earth floors for warmth and comfort, but the urine will seep into the ground, and you will lose valuable nutrients from the composted bedding. Concrete is the most popular option as it is easy to keep clean, and you can give it an even, slight slope to a drainage point in the centre of the house. The only drawback with concrete is that it is cold and hard, so you will have to insulate it adequately and give the goats extra bedding material, especially in winter. It is advisable to provide sleeping benches in a concrete-floored goat-house, to avoid any problems with rheumatism.

Walls

If you are building from scratch, bricks or breeze-blocks are the best options for the walls as they will retain the heat and give a durable structure, although timber will suffice for a small lean-to house, or one that is not too exposed to the elements. If you can incorporate hopper-style windows on the south-facing side these, along with the roof ventilators, will enable you to keep the house airy and comfortable on even the hottest day. If you paint the house white inside, this will make it seem lighter and more airy, and flies will be less of a problem.

Roof

The roof need not be so stout, but avoid using any material which will cause condensation. There should be skylights or fixed ventilators on opposite sides of the roof to provide ventilation.

Pens

When constructing the pens, remember that goats will chew, eat or smash flimsy materials. If possible, use tubular steel or reinforced stout wooden planks, and take particular care to ensure that the pen-doors are robust and secure. It is amazing how clever goats are – especially when it comes to working out how to undo door latches. It is a good idea to add a barrel bolt low down on the door, both as extra insurance and to give the structure more strength. As you will discover, goats are curious creatures and will make a point of standing on the door with their front feet to get a better view. If you provide them with some sort of stand for the purpose, it may save having to replace the doors before their time.

FURNITURE AND FITTINGS

A large-scale commercial goatkeeper will have a wide range of specialised equipment and fittings, but the following should be adequate for the amateur with a small flock.

Hayracks

Some keepers make do with haynets, but these do present problems. Goats can pull them down from their anchorage points, wasting a large part of the contents. In her eagerness to get at the hay an animal may trap a foot in the netting, and kids have been known to be strangled by these nets.

The most satisfactory method is to use hayracks. These are usually made of wood and are about 2ft (60cm) deep with bars approximately 2½in (6cm) apart. The goat pulls the hay through the gaps between the bars. If you wish, you can put a two-sided rack on the partition between two pens to save space.

Buckets

In most houses, each pen-door has a hole through which the goats can drink their water and liquid feed from buckets. These buckets should be made of plastic, for preference, and can be slotted into circular iron holders. It is not advisable to leave pails inside the pens, as they will inevitably get knocked over. Even if you secure them in holders, the contents are sure to be contaminated with bits of hay or droppings, and goats are irritatingly fussy about the quality of their water.

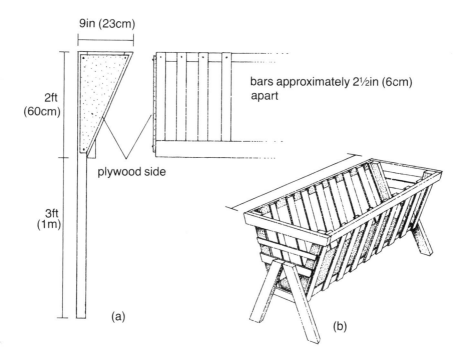

9in (23cm)

2ft
(60cm)

bars approximately 2½in (6cm)
apart

plywood side

3ft
(1m)

(a)

(b)

*(a) Wall-mounted hayrack. (b) Free-standing hay
feeder.*

Milking Bench

Unless you are planning to keep more than a few goats, a separate milking parlour will not seem worth the space or the financial investment,
and a milking bench, constructed outside the pens, is perfectly
adequate. A typical bench consists of a raised, rectangular wooden
platform, to which a reluctant goat can be tethered during milking. A
food container fastened at the head end will keep the goat happy while
you do your work. This arrangement obviates the need for a milking
stool, since you can sit on the side of the bench instead.

Milking Equipment

Milking equipment is described in Chapter 9, but you should consider
at the planning stage whether you are going to be able to incorporate a
separate dairy or whether you will have to use your kitchen for dairying tasks. The prime requisite of a dairying area is that it be completely

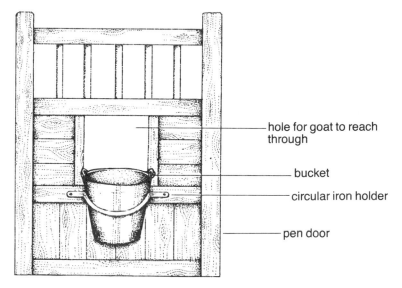

hole for goat to reach through

bucket

circular iron holder

pen door

Buckets should be secured outside the pen door or they will soon be knocked over, and their contents become contaminated.

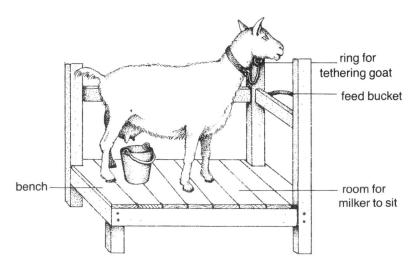

ring for tethering goat

feed bucket

bench

room for milker to sit

Milking bench with room for the milker to sit by the goat.

dust- and germ-free to avoid contamination of the milk and of sterile equipment.

Miscellaneous equipment

You will have to have a few tools for mucking out – a fork, a shovel, a yard broom and a wheelbarrow are the bare essentials – as well as other items such as hair clippers, hoof trimming tools, brushes and a medicine cabinet. Obviously, this equipment is best kept out of harm's way and the optimum arrangement would be to have a separate storage cupboard.

SERVICES

You will find life very hard if there is no supply of running water in the goat-house. Goats are thirsty creatures, drinking up to 6 gallons (27 litres) of water a day, and you will need plenty more for cleaning. Also, if your dairy is in the goat-house, running water is the best medium for cooling the milk quickly. Electricity is the other service which makes life easier for a goatkeeper, especially in the dark days of winter.

4 Management and General Care

There is no getting away from the fact that goatkeeping involves a lot of hard work and a great deal of time. The number of chores may seem daunting, but once you have established a routine you will find that you take it all in your stride. By being meticulous and methodical about seeing to the goats' needs, what could become a succession of arduous chores is transformed into a very satisfying and enjoyable enterprise.

ROUTINE

It follows from the above that the establishment of a routine is the first priority. This routine will be governed to a great extent by the goat's natural daily cycle and by the proportion of the day spent foraging outside, but you will soon discover the best routine for both yourself and your livestock.

A goatkeeper's regular tasks divide into daily, weekly, monthly and seasonal jobs. It would be a good idea to make yourself some checklists for the first few months, until you have got into the swing of things, and then later you can simply jot down reminders in your diary. Your list might include the following.

Daily

Feeding
Watering
Milking
Pasturing and exercise
Mucking out
Tidying
Dairy jobs
Grooming

Weekly

Fencing
Tending crops and grass

Monthly

Hoof trimming
Thorough cleaning-out

Seasonal

Worming
Vaccination
General repairs

Feeding and watering are covered in Chapter 5, and milking in Chapter 9, but a few words on milking management will not go amiss here.

MILKING

The two most important aspects of milking are regularity and hygiene. As with feeding, it is no good suddenly deciding that you do not feel in the mood – your goats depend on you for nearly everything. Once you and your goats have established a routine for the twice-daily milking, stick to it. Milk them in the same order every time, and at the same hours each day and they will give of their best.

As for hygiene, since goats' milk does not need pasteurising – they are less susceptible to the diseases which plague cows and which can be transmitted to humans through the milk – you will have to keep all your equipment, including your hands, scrupulously clean. Also, make sure that the milking environment is dust-free. If possible, carry out dairying activities in an isolated section of the goat-house or, if this is not possible, in your kitchen.

PASTURING AND EXERCISE

Where feasible, goats should be let out to graze. Even if there is no grazing area, they need at least a short exercise period – say a quarter of an hour – outside, unless the weather is very bad, in which case you would do better to keep them inside. It is beneficial for a goat's hooves

Anglo-Nubian kids in paved and grass exercise yards.
The paving will help to keep their hooves in shape.

to have some exercise on hard ground, as this will help to keep them in shape and avoid your having to trim the hooves too often.

You can soon tell if a goat is unhappy with the weather. She should be taken in as soon as she lies down, looks cold or appears generally disgruntled or miserable. Remember that a slight chill in a goat can easily develop into pneumonia. If the goats are drenched in a sudden shower, give them a brisk rub down to dry them off and get their circulation going.

It is a bad idea to tether your goats outside, as there is a risk of their being strangled, pestered by the unwelcome attentions of dogs or passers-by, or of suffering exposure to the elements when they would rather take shelter.

MUCKING OUT

During the winter it is advisable to leave the bedding – consisting of straw, droppings and urine – in the pens until it has built up to a comfortable depth for the goat. You should even add a little straw to the mixture from time to time. You may find that your goats pull down as much hay on to the floor as they eat, but straw is a lot less expensive, so you would do better to find some way of adapting the hayrack to reduce wastage. This system may seem unhygienic, but the composting mixture keeps the goats warm and insulates them from the cold. The smell is also surprisingly inoffensive. A good depth of litter is not, of course, so important if the goats are given sleeping benches.

The daily task in winter is to turn the bedding and in summer, when the extra warmth is not needed, to clean out any messy litter and replace with fresh straw. Whatever the weather, the floor outside the pens should be washed down, and the goats' water replaced. This is probably best done after the morning milking.

The muck heap is an inevitable by-product of goatkeeping and can form a valuable source of fertiliser for crops on a smallholding, and for the garden, friends and neighbours if you are a 'back-garden' goatkeeper. In the latter case, think hard about where you site the heap. You should not place it in such a way that the prevailing winds will waft the odour straight to your neighbours' barbecue area or back door.

TIDYING

While you are freshening up the floor and the goats' buckets, have a quick tidy round, making sure that everything is in its proper place. Tidiness and cleanliness contribute not only to hygiene but also to your enjoyment of goatkeeping.

DAIRY JOBS

You should sterilise all your dairy equipment once every day, either with a sterilising solution, in an oven, or by steaming it for five minutes or so. The best time for this is after milking. Another regular task, although it need not be done every day, is to wash down the walls and surfaces of the dairy room. You should also make a note of any problems or illnesses the goats might have, on the milk records, to help you interpret the records properly when you are making periodic assessments of yields for each goat.

If you are making cheese or butter, this is bound to entail daily chores. These will be covered in Chapter 9.

GROOMING

A final daily task is brushing your goats down. You will need a dandy brush and a fine-toothed comb for dealing with long hair, a body brush for smooth areas, a rubber curry for helping to bring out the old coat at the end of winter, and a soft cloth for 'finishing off'. Go easy with the brushes during the winter, as you must avoid dislodging the goat's soft, warm undercoat.

If you are not planning to show your animals, grooming might seem an unnecessary job, but it is the equivalent of our regular baths and has several other advantages. Brushing is an effective way of removing fleas and lice to which, like any other domestic animal, goats are susceptible. Regular treatment cleans the skin as well as the hair, so making the goats less attractive to these parasites. Similarly, it helps to get rid of scurf in spring when the old coat is coming away. In addition, a brisk brushing stimulates the animals' circulation and helps to keep them healthy.

Grooming with a dandy brush to remove surface dirt.

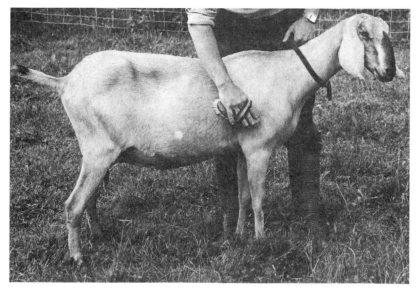

The body brush is best for smooth areas.

FENCING

Both the weekly tasks mentioned in the list above are only for those goatkeepers who have room on their land for pasturing and crop-growing. Fencing is used to restrict the grazing and browsing area of the flock. It can take various forms, but certain types of fencing are definitely unsuitable – any under 4ft (120cm) high, wire-netting fences (your goats will soon break it down to reach the juicy titbits on the other side), barbed-wire (they will only hurt themselves on it) and any flimsy structure. A living hedge may seem a suitable option, but there are few thorny bushes which a goat will not either plough or eat through, and planting a poisonous hedge is unwise since new goats may not recognise the species as poisonous.

Many goatkeepers use electric fences – particularly for temporary fencing – and these are perfectly suitable, providing that the animals are supervised until they appreciate what the fence can do to them if they try to breach it. However, if you have a male goat, he will happily endure the momentary discomfort in return for freedom. An electric fence has to have evenly spaced live wires with the top one no lower than your goats' eye level.

Angora herd confined by electric wire fencing.

*The most commonly used fence is made of stock-proof
wire, staked at regular intervals.*

The most commonly used permanent fence is made of stock-proof
netting, staked at regular intervals and strained taut. This solution
seems proof against most escape plans.

To maintain your pasture, you will not want the goats to have free
access to the whole area, and it is best to divide the paddock up into
strips, moving the flock along each week. Not only does this give the
rough grass and bushes time to recover from the goats' ministrations,
but it also helps to prevent the build-up of worm infestations. You
could, of course, have the paddock divided up permanently, but for
optimum flexibility it is better to use temporary fencing which can be
shifted around at will. This is where an electric fence scores highly – it
is infinitely mobile and adaptable. If you are going to adopt this system,
you will need to move the fence once a week to provide the goats with
fresh pasture.

If you can train your goats to respect the live wires, this can prove
extremely useful since, providing you do not let them investigate too
closely, you can use ordinary, non-electrified wire in other situations
to help guide them to where you want them to go. But beware – if

they are given the opportunity to discover that the wire is not live, they will need retraining to the electric fence.

CULTIVATING CROPS AND GRASS

Again this is a task for those who are fortunate enough to have sufficient land for cultivation and pasture. For full details, refer to Chapter 5.

HOOF TRIMMING

If your goats stay indoors most of the time, their hooves are not going to have the wear which keeps them in shape in the wild. There is a way of providing them with exercise on a hard surface, however, even when they are kept indoors. You can leave a large block of stone in the pen as a look-out stand, and this will help to maintain the condition of their hooves. Nevertheless, they will need trimming occasionally, and you should check all your goats monthly for this. If a hoof is allowed to grow too much, it can cause lameness and foot-rot.

The process of hoof trimming is straightforward enough, once you have practised a little. You can do the job either with a paring knife or with trimming shears, both of which are readily available from farm suppliers, although you may find the shears (which look rather like straight-bladed secateurs) easier to handle.

A goat's foot has an outer wall, a sole and a heel. It is the outer wall that you are going to trim. It is easier to do the job after the goat has been walking around in wet grass for a time, as this softens the horny substance. Tether the goat, then pick up one of her feet and clean out any debris with either a knife or the point of the shears. Now trim off the side walls, flush with the sole, and any excess heel and toe. Repeat the process with the other hooves and check that the goat is standing square on her feet. If the animal is rather lively or rebellious, another pair of hands may come in useful.

THOROUGH CLEANING-OUT

At least once a month in winter, and more frequently in the summer, all the pens need a thorough clean-out. Remove all the litter then wash down the walls and floor with a weak disinfectant solution. Then lay fresh straw in the pens.

Hoof trimming. Before . . .

during . . .

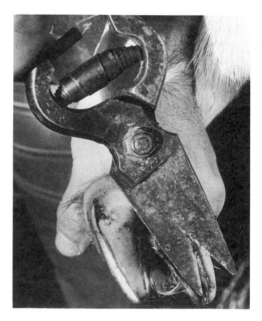

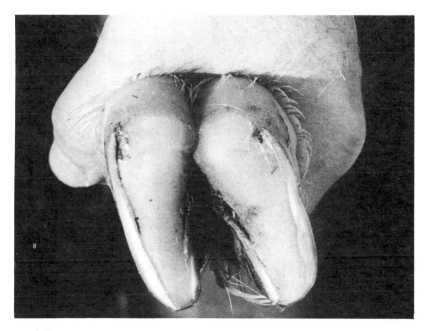

and after.

WORMING

Worms are distressingly common in goats, although there are certain measures you can take – such as the grazing rotation mentioned above – to reduce the scale of reinfestation. The worm eggs are excreted in the goat's pellets and soon hatch out. As the goat grazes, she takes in the worms with her food, and the cycle begins again. Even the scrupulously methodical and hygienic goatkeeper cannot prevent the animals becoming infested at some time or other, and worms severely affect a goat's general condition and milk yield.

Consequently, worming should be a regular routine, although the frequency varies depending on the age of the goat, the season, and whether she is in-kid or not. Kids should be wormed between twice a year and monthly, according to the scale of the worm problem, beginning when they are two to three months old. The same goes for milkers, although they should not be wormed within a month of kidding.

The most commonly used vermifuge is Thibenzole, which is available from vets. The method for administering it is painless and simple. The dose varies from one pill for a full-grown adult to half a pill or less for a kid. Dissolve the crushed dose in some water (you can also obtain Thibenzole in liquid form which makes life easier) and put it in a narrow-necked plastic bottle. Hold the goat's head close to your body and tilt it *slightly* upwards with your left arm, then insert the neck of the bottle gently into the gap at the side of her mouth. Pour the liquid slowly into her mouth, stopping immediately if she shows signs of choking. This process is called drenching. The crushed tablet can also be added to the animal's food, or administered whole with a handy implement called a pill gun.

As with all such treatments, the worms may well build up some resistance to the vermifuge if you use the same kind every time, so it would be as well to give the occasional treatment with a different drug, such as Nilverm, Nemicide, or Panacur which is probably the best drug for this purpose as it is extremely effective yet very safe.

You should also check regularly for lung worm and fluke, neither of which is very common, but both of which can prove harmful. Fluke, which is only present in marshy areas, where it uses the snail as a host, can prove fatal.

Drenching.

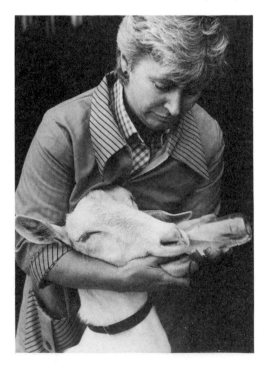

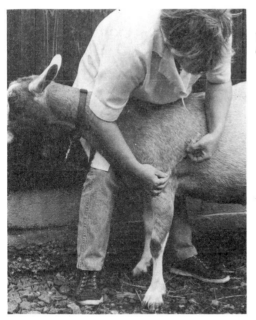

Vaccinating against entero-toxaemia.

VACCINATION

As will be discussed in Chapter 7, one of the most deadly diseases in goats is entero-toxaemia (literally, 'stomach poisoning'). It can kill an apparently healthy goat within twenty-four hours and, although one fatality acts as a warning to inoculate the rest of the herd, it is better to have all the goats inoculated routinely, twice a year. The vaccination will protect the animals against all the *Clostridii* organisms. In the first couple of years it would be best to ask your vet to do the vaccinations, but it is not a complicated procedure and you should soon feel confident enough to do the job yourself.

GENERAL REPAIRS

No structure will last forever, especially when goats are around, so you will need to spend time in the spring and before the onset of winter making sure that all your perimeter fences, pens, and the goat-house itself, are in good order.

5 Feeding

THE GOAT'S DIGESTIVE SYSTEM

The goat is a ruminating animal and has the digestive system typical of this group. Like cows and sheep, goats have four stomachs – the rumen, reticulum, psalterium and abomasum, in order of food progression. You may wonder why any animal needs four stomachs, but in fact this is an extremely efficient system for an animal which may have to snatch its food in bulk in a short time.

The principle is that the food goes straight from the mouth after brief mastication to the rumen and reticulum where it is partially digested, after which the coarser material is regurgitated (the 'chewing the cud' stage), followed by passage into the psalterium and abomasum and thence into the intestines. If you watch your goats closely, you will see that they use two different techniques for swallowing. When they first take in food, they give it a perfunctory chew to soften and shape it into a manageable mouthful, then draw in their necks to let the food slip into the rumen. After they have finished the lengthy process of chewing the cud, however, they stretch out their necks so that the cud can pass straight into the psalterium and abomasum.

It was probably the life-style of these animals in the wild which led to the evolution of this digestive system – they could bolt down their food while there were no preying enemies around, then retreat to a safer spot to chew in peace. The rumination process is also dictated to some extent by the dietary needs of goats. Since they feed only on vegetable matter which is made up largely of carbohydrates, including cellulose and water, they need this special process to deal with their food.

BASIC DIET

A goat needs some protein and a very large amount of carbohydrate and starch, with the correct mix of vitamins and minerals if she is to stay healthy. A large proportion of her food should consist of fibrous material.

In the wild, a goat will automatically search out fibrous plants and tree bark, selecting only the best quality food. Unfortunately, if you rely solely on grass, leaves, greenstuffs and root crops, the goats will not be getting enough protein or vitamins and minerals, especially for milk production, and this is why they are usually fed concentrates and mineral supplements.

Hay

The uninitiated may think that hay is simply dried grass and that if you grow your 'weed-free' lawn grass, harvest it, dry it and feed it to your goats you will be doing them a great favour. This is not true. If you were to offer your goats hay made from the finest grass seed, they would either turn their noses up at it or, if they did deign to eat it, would soon be decidedly out of condition.

The best hay for goats consists of a good rough mixture of leafy grasses and other plants, such as red clover, nettle, lucerne, and even pea straw and cereal straw to provide extra roughage. The process of digesting all this coarse material generates quite a bit of heat and can insulate your animals from the cold if they have a good feed of hay before venturing outdoors. It pays, therefore, to buy the best hay you can afford.

Greenstuffs

Cabbage, sprout tops, broccoli, cauliflower leaves and kale are very popular with goats, although you should avoid feeding goats kale for at least four hours before milking in case the flavour taints the milk.

Roots

In the wild, goats rely on root crops to a great extent to carry them through the winter. In the domestic environment, the most frequently used root crops are mangolds, swedes, carrots, turnips and fodder beet. Always cut root vegetables up into manageable portions. Again, the stronger vegetables should not be fed prior to milking as their flavour may pass into the milk.

Trees, Shrubs and Wild Plants

As some goatkeepers will know to their cost, goats love tree bark. Fortunately, though, they will be just as happy with hedge trimmings and branches of elm, chestnut, willow, ash or apple trees. Many wild

Feeding

Anglo-Nubian, browsing on willow.

plants are perfectly safe, but there are quite a few that are poisonous (*see* pages 80–82).

Minerals

Although goats will obtain a certain amount of minerals from their normal food, it is advisable to give them an additional supply, either in the form of a supplement mixed in with their concentrates – if these are home-produced – or by means of a mineral lick brick which can be hung up in the pen.

Concentrates

Concentrates are usually made up of various mixtures of crushed, flaked or rolled high-carbohydrate grains, such as oats, maize, barley and bran, with other higher-protein foods – decorticated groundnut cake, locust beans, linseed cake, soya bean meal, sunflower seed cake, and so on. Sugar-beet pulp is a must for milkers as it promotes high yields. However, it should not be fed dry since it absorbs a large amount of liquid and can lead to bloat – soak it in hot water first.

The proportions and quantities of concentrate mixtures vary depending on the condition, age and function of the goat, with the milker requiring most to maintain or increase her yield. Many goat-keepers like to make up their own concentrates, and the various permutations and compositions could fill a book just by themselves. The novice would probably be better advised to buy ready mixed concentrates to begin with, and to experiment with homeproduced mixtures once he or she is better acquainted with the theory – and with the goats!

Miscellaneous

Goats love sliced apples and a variety of kitchen and garden waste such as potato peelings (especially when dried out in the oven), pea pods, dry bread, prunings, and so on.

Water

Water is a vital part of any animal's diet, and a goat can drink vast quantities – up to six gallons (27 litres) a day. If you do not leave water with the goats, you should offer it to them at least four times a day. Some animals enjoy a little salt in their water, or a sprinkling of oatmeal, and all goats seem to drink water more readily if it has been warmed.

FEEDING ROUTINE

There are no hard-and-fast rules about what should be fed when to your goats; the most important aspect of feeding is that they should have a balanced diet and one that is suited to their needs. For example, milkers will benefit from a higher proportion of protein in their diet than other goats but if a milker's rations were fed to a maiden goatling, she would soon become fat. One cardinal rule, however, is that you should never make an abrupt change in your animals' diet, especially to much richer food, as this can cause untold health problems.

A typical daily feeding timetable might consist of giving half the day's ration of concentrates at the morning milking, followed by hay. Later in the morning the goats may go out if the weather is suitable to graze and forage. If they have to stay inside, they should have plenty of greenstuffs and some root vegetables or branches. At the evening milking they will have the rest of their concentrate and the hayracks will be replenished.

As for quantities, a daily maintenance ration for a goat should consist of approximately 1lb (450g) of concentrates, 5lb (2.25kg) of hay, and 10lb (4.5kg) of greens, branches or roots. A milker will need

Feeding concentrate after the evening milking.

extra concentrates to give her the added protein – approximately 1lb (450g) more. These figures are only averages and will, of course, vary according to the amount of milk the goat is giving.

HOME-GROWN CROPS

Although you could purchase all the food for your goats, this would involve considerable expense, even if you are able to arrange with a friendly greengrocer or market stallholder to buy his unsold produce. You do not need that much space to grow at least some greens and root vegetables, and, if you have only two goats, such a course could have some impact on your feed bill. As a bonus, you can make sure that they are top-quality vegetables, produced without recourse to chemicals – the most useful crops would be kale, cabbages, maize, fodder beet, mangolds, carrots, swedes, comfrey and lucerne.

You will want to store your root crops through the winter, and the easiest solution is to make a clamp using insulating plant material (such as bracken, heather or straw) as a base, and covering the roots with plenty of earth.

6 Breeding and Care of Young

One of the great pleasures of goatkeeping, without a doubt, is assisting at the birth of kids and having the satisfaction of seeing them grow up healthy and content, thanks to your management.

AGE FOR MATING

Most kids are born in the spring, and the rutting season usually begins in the autumn. It is best to wait until your goatling is in her second year before mating her so that she can establish herself and reach full strength before the first kidding. Provided that a goatling is well grown, she can be mated at fourteen months; it all depends on her size. If you have two goats, the normal practice is to mate them in alternate years, so that you have a continuous supply of milk. As most goats can breed until they are at least seven or eight years old, holding back the first kidding until the second year still gives plenty of scope for breeding.

OESTRUS

The female goat's cycle lasts around twenty-one days, with an oestrus, or 'heat', period of approximately two to three days. The cycle continues from September until February. When a goat is on heat she will be unusually restless and vocal, and her tail will wag vigorously. The physical signs are a swollen vulva and a mucus discharge. Milk yield will either increase or decrease, but whatever the change, it will be marked. Confirming oestrus is another use for milk records.

CHOOSING A STUD GOAT

Once you have decided that you are going to mate one of your goats, you will have to find a male. I am assuming that you will not, to begin

Saanen male goat.

with at least, have your own male goat; indeed I would not advise keeping a male if yours is only a small-scale operation, as they require separate management and consequently a great deal of time.

Even if you are not interested in showing your animals, it still pays to select a high-quality stud goat. All the breeds have been developed with high milk yields in mind, and the male of the species carries these genes as well as the female. It is not advisable to mix breeds. If you keep the breed line pure, your first kids might include a future Breed Champion – this is just one more exciting aspect of breeding.

There are several ways of finding a stud goat. You would do well to begin by looking in your own area, since this will reduce the distance you have to travel – your local club's newsletter is bound to include details of stud goats in your region.

Once you have chosen your stud, plan ahead to ensure that you will have transport, as it is your goat that will be visiting the male, rather than the other way round. Contact the owner well in advance to let him know that you want your goat served and to discuss arrangements for the mating. The normal practice is for you to contact the owner on the morning that your goat comes in season so that you can agree your arrival time. The law requires that you keep a record of your stock movements. This can be a simple notebook where you note down the date, name of the goat, location at the start of the journey and the destination.

THE SERVICE

The male serves the goat twice in one visit, so that there is a greater chance of the service 'holding'. Before you leave, you will pay the owner his fee and he will give you a service certificate. If, three weeks later, the goat comes in season again, the owner will usually allow her to be served again, free of charge. If this second mating does not take, you should consult your vet, as the goat may be infertile. Assuming that the mating is successful, the kids will be born five months (145 to 155 days) later.

MANAGEMENT OF THE IN-KID GOAT

Make sure that the mother-to-be has a good and varied diet. Her milk production may be drying up, but she needs the same amount of food for kid production. She also needs a reasonable amount of exercise, and this should be supervised, since any kind of fright or accident could easily lead to an abortion. Expectant mother goats tend to laziness, so you may well have to take her for a walk.

Take special care with her diet in the last six weeks of pregnancy, as this is when the kids put on most of their weight, and she will soon need a lot of energy for milk production. She should therefore be given plenty of concentrates at this time.

PREPARING FOR THE BIRTH

About two weeks before kidding is due, scrub out and disinfect the goat's pen (or a separate kidding pen, if one is available), lay fresh straw and let her loose in the pen. Make sure that you are familiar with her hindquarters long before the due date for the birth so that you can recognise the changes that occur as kidding time approaches – the bones around her tail will slacken to make room for the birth and her vulva will look swollen. Also, just before birth, she will become fidgety

55

and her udder will be full and tight, ready to feed the new arrivals. It may be that her udder fills a week or more before kidding, in which case it is probably best to take some of the milk to relieve the pressure. Do not milk her dry, however, as you risk bringing on milk fever after the birth (*see* page 76). Keep an eye out for signs of pregnancy toxaemia (*see* page 72) in the last few weeks.

SIGNS OF IMPENDING BIRTH

These vary from goat to goat, but it is common for her to become much more noisy and to paw at the bedding. She will seem to have lost weight as the kids move down, and her breathing will become quicker. There will be a mucus discharge from the vulva. She may strain and grunt, but this is not a cast-iron sign that anything is about to happen. Just watch and wait.

THE BIRTH

Most goatkeepers are apprehensive about their first kidding but, more often than not, there is absolutely nothing to worry about. The goat will probably, but not necessarily, lie down to give birth. The normal presentation of a kid is head first, with its nose resting on its two forelegs, and passage of the head and legs in this position should be trouble-free. The appearance of the head may be preceded by the bag, or the kid may still be inside the bag. In either case this will burst, lubricating the delivery. The rest of the kid should appear in short order. If the proceedings seem to have ground to a halt, call your vet. The whole birth process takes a matter of minutes rather than hours, and is usually accomplished with four or five contractions.

As soon as the kid is born, its mother will lick it down. The kid will probably sneeze. This helps to clear mucus from the air passages and to encourage breathing. The umbilical cord usually breaks by itself but if it fails to do so, tie it off a few inches from the kid's body, cut it on the mother's side of the knot and paint it with iodine tincture.

If everything goes smoothly, the second kid should arrive after a few minutes, and there might even be another. If you are unsure whether a third kid is waiting in line, feel the goat's belly, just in front of the udder. If there is a hard lump, a third kid is on the way.

The afterbirth should be delivered within hours of the kids' arrival. If it is not, even after you have lifted up her belly by way of encouragement, consult the vet, since a retained afterbirth can cause problems.

A newly born kid.

When the proceedings are over, wash the mother down with warm water, dry her off, change her straw, refill her hayrack with the best hay you have, give her a warm drink containing treacle or glucose, and provide her with some greens. Having made sure that the kids are sucking properly, leave the proud mother to have a well-deserved rest. It is very important that the kids drink an adequate amount of this first milk as it contains all the antibodies which will provide them with a natural immunity to any bacteria present in the goat-house.

MULTIPLE BIRTHS AND COMPLICATIONS

Most goats have two kids, although one, three, or even four kids are not uncommon. It is unlikely, in the case of triplets or quadruplets, that all the births will be normal presentations, but abnormal ones can easily be remedied. It is surprising how Nature manages to cope with most problems, and I would advise anyone not to interfere unless it is absolutely necessary.

You may not feel confident enough to correct abnormal presentations at your first kidding, and it may be wise to call in an experienced goatkeeper or your vet to assist until you have enough practice. Assisted births usually require antibiotic treatment to prevent infection, so you would probably be well advised to call in the vet anyway.

If you are going to put your hand inside the goat, first wash the goat's rear, then scrub your hands and arms clean and apply some obstetric cream. Make sure that your fingernails are short and clean and keep your fingers together inside the animal. You will need an extra pair of hands to hold the goat while you are working inside her.

If a kid presents rear end first, with its hind legs tucked under, push it back into the uterus and bring the legs towards you – the kid will come out backwards. If the front legs are presented but the head is not visible, it is probably twisted and you should push the kid back, gently turning its head into the normal, forward-facing, nose-down position. Occasionally one of the front legs appears, while the other stays tucked back inside. As before, push the kid back, then straighten the legs into the normal position.

There can be other problems. In some multiple births the kids and umbilical cords get all mixed up inside. In this case, it is a matter of getting in there and sorting them out – a task for the experienced goatkeeper only. In very rare cases a kid dies *in utero,* and you will need professional help.

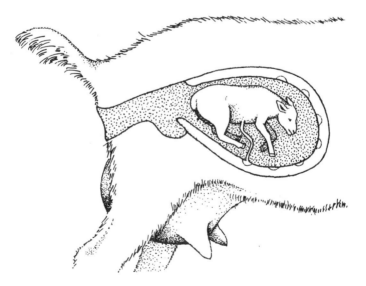

A breech – the kid presents rear end first and will come out backwards.

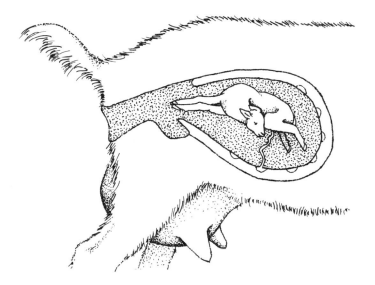

Head back – push the kid back and bring the head forward.

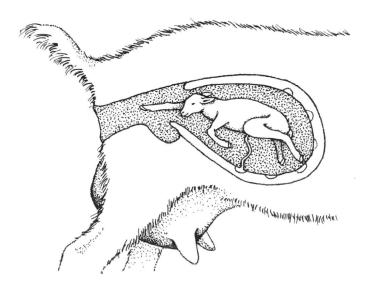

Leg back – straighten the legs into the normal position.

WHEN TO CALL THE VET

It is difficult to give absolute advice here. All I can suggest is that if you are worried, call him. Goats are valuable animals, and it is better to be too cautious and feel embarrassed at having called the vet in unnecessarily, than to be wise after the event. Nevertheless, you will be reassured to learn that most kiddings are trouble-free.

AFTERCARE

The Mother

Do not milk out the goat for at least four days after kidding to avoid the risk of milk fever. You may find that there is a considerable reddish discharge from her vulva for a while, but this will stop within two to three weeks. It is perfectly normal and need not give cause for concern, although you should clean her rear end up periodically. Feed the mother only limited amounts of concentrate for the first few days, bringing her back up to the full ration very gradually.

The Kids

Whether you eventually decide to let the kids take all their milk from their mother or not, it is best to keep the family intact for the first four days. During this time you should decide what is to become of the kids. You will have to let your head rule your heart, for unless you want to rear a buck, there is little else to be done with a male kid than to have him put to sleep. If you should decide that you want to keep a male, basic management is outlined at the end of this chapter. Never give a male kid away as a pet – it would just be cruel. Within six months he will be big, boisterous and smelly, and may well be passed from family to family until he ends up a neglected, cantankerous, totally unlovable animal, and is eventually destroyed.

Female kids should be examined thoroughly for defects. Hermaphrodites can usually be distinguished by an enlarged clitoris and a rudimentary male sex organ within the lips of the vulva. Supernumerary teats (extra teats), double teats (where a teat has two openings), unsound legs, crooked backs, twisted faces or malformed jaws are other serious defects. You will probably decide to have any such afflicted kids destroyed. Bent ears are usually no problem as they straighten themselves out.

REARING

The Choice

At the end of the kids' four days with the mother, you will have to make up your mind whether those kids you have decided to rear will be suckled or hand-reared.

The more natural way may seem attractive, but it is going to be impossible to measure the mother's milk yield accurately. Also, the kids may not feed equally off both teats and this may result in lopsided udder development. The advantages of hand-rearing are that you will be able to monitor the mother's milk yield, and the kids will grow up to be much more manageable and less timid.

Bottle-Feeding

Most goats that are hand-reared are bottle-fed, as this is much closer to the natural method and ensures that the milk will not be drunk too quickly. (The latter can result in poor development of the digestive system.) If you are going to hand-rear your kids, they should be taken out of the pen, in their mother's absence, four days after birth. If you give them an adjoining pen there will be less chance of the mother being upset.

The utensils for bottle-feeding must be kept meticulously clean and the milk must be fed at blood temperature. A baby's bottle is perfectly satisfactory for feeding kids and, if it has a wide top, you may milk direct from the mother into the feeding bottle, after rinsing it out with hot water. When you are feeding the kids, make sure that the teat does not come away, as swallowing it will not do them any good at all!

Start with four feeds a day and let them take as much as they want. This may not seem much in the early days, but they will soon increase their intake until, at about four weeks old, they will be drinking about one pint (half a litre) at a time. By this stage, you will have moved them on to a much larger bottle with a robust calf teat and, you can reduce the number of feeds to three a day. At six months, they will be having just one bottle-feed daily. As for what kind of milk you give the kids, you can either feed them straight from the mother or gradually change over to milk substitute after about four weeks.

The kids can start eating cereals after their first week. They will only take a little to begin with, but their protein requirements are being supplied by their milk and pasturing, so you should not be concerned. They can also be introduced to hay and dry greens at this stage. They should be encouraged to graze from about two weeks. Gradually, over

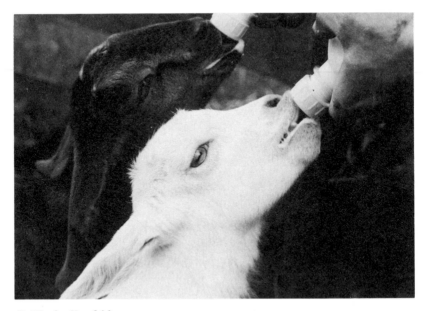

Bottle-feeding kids.

the first few months, introduce new foods until they are eating a fully varied diet. Once they are weaned, they should be offered water regularly.

By their first birthday, when they become goatlings, they will be having an adult diet, but you should not give them as much protein as a milker or they will become overweight.

Management

Kids and goatlings require plenty of exercise. If you can provide something for them to clamber about on and jump off this will help them to keep fit and develop well. Like adults, they should have their hooves trimmed monthly and be wormed at regular intervals. Handle your young animals as much as you can and, if you train them to wear a collar and walk to a chain or lead from the age of two months, you will save yourself some problems later on.

First lessons on lead training.

DISBUDDING

One of the checks you should make on newborn kids is whether they are horned or hornless. A horned kid will have a swirl of hair over the embryonic horns. Whatever their sex, kids should be disbudded, and as early as possible. The law requires that this task be carried out by your vet.

THE MALE GOAT

Male goats have barely received a mention so far in this book. This is because most people want to keep goats for their milk, and a male will serve little purpose in the economy of their goatkeeping operation. However, if you should wish to rear your own male, a few words on his management may be helpful.

The first consideration is that a male has to be kept apart from the rest of the herd. A male kid can serve a female as early as three months of age, so the separation must be carried out fairly early on. He should be housed in a completely separate part of the goat-house, as his powerful smell is liable to taint the milk. Indeed, you should always don a plastic mac and gloves before entering the male's house, to avoid transporting this odour into the milkers' section.

The diet for a male is similar to that for a female, except that, because of his superior bodyweight, he will need more of everything. Omit dried sugar beet pulp and mangolds from his diet, as these can block the urinary tract. The same is true of hard water, so you would do best to offer him rain water, if possible, rather than tap water.

Never let children play with a male kid, as this becomes dangerous once the animal is adult, and it will be hard to eradicate 'playfulness' from the male's personality. Let him out for regular exercise in his own yard, treat him firmly but kindly, and you should have no problems. If he can see the other goats he will be content.

Routine hoof-trimming and worming will be necessary, as with females, even though these can be eventful and slightly unpleasant duties.

Service

A healthy adult male can service up to a hundred females in a season with a maximum of three per day, although in his first year the total number should be restricted to thirty.

7 Accidents and Diseases

If you are thinking of keeping goats, this chapter might well put you off for life as you envisage your precious animals dropping like flies around you with a host of frightening ailments. But a lot of illness in goats can be avoided by adhering to good management techniques – giving a sensible diet and taking common-sense hygiene precautions, especially where kids are concerned.

DIAGNOSIS

Signs that all is not well with one of your animals include poor coat condition or appetite, lack of alertness, a change in the droppings, alteration in milk yield, bloatedness, a lack of cud chewing, lameness, watery eyes, coughing, signs of anaemia and evidence of pain, such as grinding the teeth. Temperature, pulse rate and breathing rate are all useful aids to diagnosis of the problem.

Temperature

To take a goat's temperature, smear the tip of a thermometer with grease and introduce it into the back passage, using a rotating movement. Hold it in place for one minute. The normal temperature lies between 39.2°C (102.5°F) and 39.4°C (103°F). A higher temperature points to a fever and a lower one to a condition verging on collapse. In either case, call the vet.

Pulse

The best way to take her pulse is to lay your fingers high up on her flank or just under the jaw where main arteries pass over bone. Seventy to eighty is a normal pulse rate, but it can be as high as one hundred and twenty in a fevered animal. If the pulse is irregular, this may be a sign of heart problems.

Taking a goat's temperature.

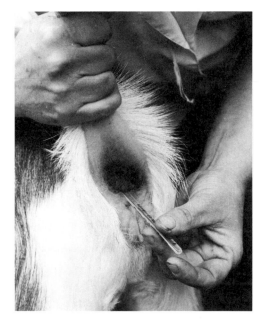

Respiration

To time a goat's breathing rate, watch her flank while she is lying down. If it is between twenty-two and twenty-six per minute, there is no problem. Abnormal respiration can indicate congestion or pain.

SHOULD I CALL THE VET?

In the early days, there will be few illnesses that you can diagnose with any certainty, so always call in the vet if in doubt. You should be able to deal with minor abrasions and accidents, and you may have a friend who is an experienced goatkeeper who could advise you, but you should not delay too long before obtaining professional advice, since some very unpleasant diseases can develop extremely rapidly.

MANAGEMENT OF A SICK ANIMAL

If one of your goats seems off colour, she should be kept inside in the warm. She will benefit from wearing a goat rug and from being in a cosy, draught-free pen. Let her have hay, but withhold concentrates until the problem has been diagnosed and treatment has been prescribed.

Unless the vet insists on isolation, she will be much happier if she can watch the daily activities of the other goats. However, she will probably enjoy a little peace and quiet when they go out for their exercise or to pasture.

Goat Rugs

Goat rugs are easy to make from old blankets, coats or any warm, non-slip material. To ensure that the rug fits, measure the goat from neck to tail and from the withers to the top of the front leg. If in doubt, cut it on the large side, tack the seams and try it on her for size. When you have achieved a reasonable fit, bind the edges with fabric tape. You can use the same tape for the ties.

You should not wait until a goat is feeling poorly before making her a rug or two – the last thing she will feel like doing is acting as a

Anglo-Nubian milker in a light-weight wool rug, suitable for a sick goat, shows or just warmth.

dressmaker's dummy! If you want an extra warm rug for the winter, make a two-layer version by lining it with blanket material.

Feeding

It will not matter if she is off her food for a couple of days, but beyond that she must be encouraged to eat. Try some choice delicacies – bramble shoots might do the trick – or offer her a bran mash made by adding boiling water or milk to a few handfuls of bran, mixed with a little treacle or honey if she likes sweet things, and leaving it to cool to blood temperature. Gruel can be tried if the bran mash is not acceptable, or to vary her diet. Simply sprinkle oatmeal into a saucepan of boiling water and stir until it is thick.

One of the most trusted 'natural' remedies is garlic – a wonderful agent for flushing toxins out of the system and particularly useful in treating pneumonia, rheumatism, bronchial congestion, mastitis and entero-toxaemia. A slight excess of garlic may produce temporary scour, but this merely shows that it is doing its job properly.

Always keep some coarse hay on offer for when she feels up to taking normal food and provide plenty of water. She might also want some milk.

Pill guns help to ensure that the tablet is swallowed.

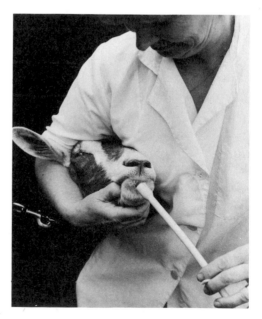

Administering Medicines

Most medicaments are administered in pill or injection form nowadays. You should let the vet or an experienced goatkeeper give the injections until you have learned the technique, but thereafter you should be able to do the job yourself. With pills, you can either use a pill gun, or crush them to be taken with food.

Some medicine-chest remedies will involve drenching (*see* page 46). Always be very careful not to pour the liquid into the goat's mouth too quickly or to tilt the bottle up too far, since in the first place, the fluid will go into the rumen, where it will sit, and in the second, you risk liquid reaching the goat's lungs which could lead to pneumonia.

THE MEDICINE CHEST

You would be well advised to keep a supply of basic medicines and first-aid equipment in the goat-house. Below is a list of recommended items, but you will make your own additions as time goes on.

Disinfectant (Savlon or similar)
Gauze
Cotton wool
Crêpe bandage
Large plasters or plaster strips
Thermometer
Rumen Stimulant
Garlic tablets
Bowl
Surgical scissors
Drenching bottle
Worming preparations
Measuring glass
Indigestion treatment
Linseed oil
Iodine tincture
Enema equipment
Liquid paraffin
Liniment
Eye ointment or drops
Epsom salts
Bicarbonate of soda

Udder salve
Flowers of sulphur
Terramycin aerosol spray
Disposable hypodermic syringes and needles
Wound Powder

ACCIDENTS

Cuts

The most common cause of cuts is barbed wire, and this is the main reason why barbed wire should not be used in fencing. Thorny hedging can also be responsible for minor cuts.

If the cut is a minor one, trim away all the surrounding hair, wash the area with clean, warm, salty water and dust with wound powder or spray with a Terramycin aerosol spray. Bathe again when necessary with warm salty water. In the summer months it may be necessary to

Wound powder, for use on minor cuts and abrasions.

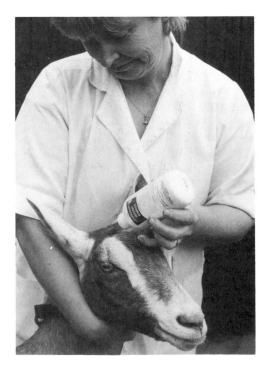

apply a fly-repellant around the wound. A vet should be called in if stitching is needed, and he may want to give the goat an anti-tetanus injection.

Broken Limbs

These are very rare in goats, but if the worst does happen, keep the injured animal still until the vet arrives.

BREEDING ABNORMALITIES AND DISORDERS

Abortion

This is hardly ever due to disease. The more likely reasons are fighting, a bad shock, over-exertion, a worming treatment given too near to kidding, or eating frozen roots. Proceed as though the goat had kidded normally, and call in the vet if – as frequently happens – the afterbirth is retained.

Acetonaemia

Symptoms: sluggishness and depression shortly after kidding; milk, breath and urine smell of nail varnish.

The disease is due to a metabolic disorder which develops gradually during late pregnancy and which is brought to a head by the sudden increase in milk production. Veterinary treatment usually includes a course of injections. Self-help treatments include a three-day course of a drench of 4fl oz (110ml) of glycerine in warm water, diluted treacle (for glucose), and starchy foods.

Cloudburst (False Pregnancy)

Symptoms: distension as though goat is in kid; discharge of large quantities of watery fluid on approximate due date for kidding.

This condition is most common in goats which are not mated each year. It does not seem to affect the goat's health and should not be a cause for alarm, although cloudburst may recur at a later date. It is best to leave well alone.

Infertility

Causes in female: malformed sexual organs; dysfunctions in the sexual organs leading to absence or irregularity of oestrus cycle; infections, sometimes indicated by discharge.

Causes in male: old age; abnormality in sexual organs (for example, undescended testicles); exhaustion; impotence; low sperm count; damage to genital tract.

Some of these root causes will respond to treatment in time, but it is generally not worth spending valuable time and energy on diagnosing where the problem lies.

Metritis

Symptoms: a thick, pink, odorous discharge from the vulva immediately after kidding.

Metritis is an inflammation of the womb which occurs either when the afterbirth has not been delivered promptly, or after an assisted birth. Your vet will prescribe a course of antibiotic injections.

Pregnancy Toxaemia

Symptoms: sluggishness; depression; tendency to stagger and, in advanced stage, to become comatose.

Like acetonaemia, this is a metabolic disorder, but it occurs before rather than after kidding. Preventive measures consist of giving the animal adequate exercise and plenty of carbohydrates and greenstuff. If you suspect this disease, call in the vet as he may have to bring on kidding to save the mother goat.

OTHER DISORDERS

Anaemia

Symptoms: gradual decline; sometimes white genital discharge.

Anaemia in goats is often caused by heavy worm infestations or by a lack of iron in the diet. Add iron to the diet (seaweed meal is useful), give a course of cobalt salt, and drench for worms.

Bloat (Blown or Hoven)

Symptoms: flanks very distended, especially on the left side where the skin can be as tight as a drum; the animal is obviously in pain.

Bloat is a severe form of indigestion, often brought on by eating too much rich food. Gases accumulate in the stomach and the animal can find no relief. The condition can be dangerous, so action should be taken quickly. Drench with an indigestion fluid or linseed oil, or try a dose of brewer's yeast, then walk the goat round and massage the body vigorously. If this does not bring relief within an hour or so, call in the vet who may have to puncture the flank.

Caprine Arthritis Encephalitis (CAE)

CAE is a comparatively new disease to this country, although it has been in evidence in the USA and elsewhere for some time. It is one of a group of viruses known as retroviruses. Adult goats, older kids and goatlings may show arthritic symptoms with stiff and enlarged joints. Younger kids show signs of encephalitis and may become uncoordinated and paralysed. Kids may be infected with the virus when fed the milk from a female which carries the disease, and it can be transmitted from goat to goat during mating. Goats may also be infected when coming into contact with other goats at shows.

There is no cure for the disease, nor is there a preventive vaccine. Diagnosis is possible by bloodtesting each goat; goats carrying the disease will show antibodies in the blood, and they are then called positive reactors. Eighty per cent of the goats tested in the USA are positive reactors. As prevention is the only cure, unless these reactors can be permanently and completely isolated from the rest of the herd, they should be culled.

The British Goat Society recommends that herds have three tests at six-monthly intervals and, if clear, yearly tests thereafter. Once clear, it is best to take various precautions to keep it that way. Larger shows offer separate accommodation for CAE-free herds. Care should be taken when visiting stud males, and to be on the safe side a male from a CAE-free herd should be used. Use a CAE-free herd when boarding goats to go on holiday and similarly offer goat-boarding to clean goats only. Goats bought in as additions to a herd should be obtained from a CAE-free herd.

The British Goat Society is running a CAE monitoring scheme — details of which may be obtained from their office.

Taking blood to test for CAE.

Chills

Symptoms: chattering teeth; scouring; the goat looks down in the mouth.

Chills can be caused by injudicious exposure to bad weather or by eating frosted grass on an empty stomach. Wrap the goat up warmly with blankets and hot water bottles and feed her a drink of warm milk or warm water and glucose plus some hay.

Colic

Symptoms: restlessness and obvious severe discomfort.

The causes of colic vary although the condition is the same – acute stomach pains. Among the possible culprits are poisoning, violent exercise after a milk feed, an intestinal blockage and eating too much concentrate.

A dose of liquid paraffin or a linseed oil drench may alleviate the problem, as might a soap and water enema. I advise you to try the first two options before thinking of an enema, since this is an energetic operation when you are dealing with a frightened animal that is also in pain.

Contagious Dermatitis (Orf)

Symptoms: very bad sores around the body orifices, including mouth and nose, which can make eating impossible and suckling kids excruciatingly painful.

This is a virulently infectious disease which appears where goats and sheep graze together, or it can even arise from very brief contact. Isolate the affected animals and either spray the sore parts with Terramycin or wash them down with salt water followed by a dab of iodine. Consult the vet concerning vaccination.

Coughs

If several of your goats are coughing persistently, you should contact the vet. It may be a symptom of parasitic bronchitis, and pneumonia could develop.

Entero-Toxaemia

Symptoms: severe scouring; constant straining; loss of co-ordination; possible fits and coma, followed by death.

As the symptoms suggest, this is the most serious disease of goats. In fact, the progression from the initial symptoms to death can take as little as twenty-four hours, so routine vaccination makes sense (*see* page 47).

Entero-toxaemia can occur so suddenly that the goat may be dead before you notice that anything is wrong. The organism responsible is *Clostridium velchii type D,* which normally sits quietly and inoffensively in the intestines, but can be triggered into vicious action, either by a change to a rich diet or even by stress due to travelling, if the goat is not used to it. As soon as you notice the symptoms call in the vet as a matter of urgency. He may not arrive in time to save the goat, but at least he can inoculate the rest of the herd.

Foot-Rot

Symptoms: inflammation which spreads under the horn of the hoof, so that the horn separates from the skin; lameness.

However religiously you trim your goats' feet, foot-rot can still occur. True bacterial foot-rot is an infection which spreads in pasture and which can be controlled by regular trimming and pasture rotation. Where the horn has separated, it must be cut away, since the gap is an ideal breeding ground for the infection. Any pockets of debris or mud should be washed out, and each hoof must be dipped into a bath or container of disinfectant or a weak Formalin solution. Keep the goats away from the infected paddock for at least two weeks. A Clovax inoculation should prevent recurrence of the disease.

Common foot-rot, caused more by neglect when mud and dung get trapped under untrimmed horn, can be treated in the same way, but is not infectious.

Goat Pox

Symptoms: small red spotty areas on the udder and teats which may burst and form crusty scabs; the affected skin is very sensitive; the goat may have a slight fever and be distressed.

This is a contagious viral disease – the virus is found in the crust. You must be scrupulous about hygiene when treating the animal. To relieve the irritation you can mix some bicarbonate of soda with water to a paste and gently spread it on the affected area. If the disease advances to the crusty stage mastitis may set in, so you should apply an antibiotic cream to the teats and the skin round them in the early stages.

Mastitis

Symptoms: inflamed, swollen udder; clots or strings in milk, plus tainting, discoloration or blood; high temperature; loss of appetite; absence of cudding; gangrene in udder in severest cases, possibly leading to death.

Mastitis is usually the result of a bacterial infection, transmitted by flies and encouraged by poor hygiene. If you clean the teat on the affected side carefully, before milking, and draw off the foremilk into a sterilised bottle, this sample, marked with your name and address and the goat's name, can be given to the vet who will have tests carried out at a laboratory to identify the bacterium responsible. He will then be able to prescribe a suitable antibiotic for intramammary injection. In the meantime, you could apply a hot poultice to reduce the swelling.

If you suspect mastitis in one of your goats, always milk her last to avoid spreading any infection.

Milk Fever

Symptoms: collapse of a milker; twisted neck; convulsions; lack of appetite; *no* fever.

When a goat is producing large quantities of milk, the strain on her

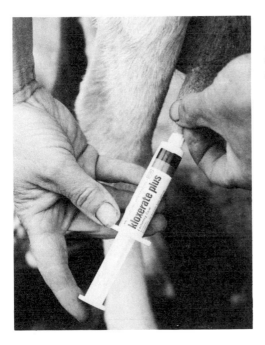

Mastitis control – giving an antibiotic intramammary injection.

system is very considerable. The cause of milk fever is a lack of calcium. Call in the vet immediately, and he will give an intravenous calcium injection. Do not let the goat lie down. If the injection is given in time, the goat will suffer no ill effects.

Pink Eye

Symptoms: the eye clouds and exudes a yellowish pus, causing severe irritation.

The condition is caused by damage to the eye from dust, flies, thorns or foreign bodies. Bathe the eye in a weak solution of salt water and apply pink eye ointment. If the eye does not respond quickly consult your vet, as neglect can result in blindness.

Scouring (Diarrhoea)

Scouring can have many causes, but it is often Nature's way of flushing a bug or poison out of the system. If at all possible, do not administer kaolin in an attempt to 'bung' up the goat, but give her plenty of water and some glucose and wait to see whether the scouring is self-limiting.

Do not let her go too long without treatment, especially if the patient is a kid, as prolonged scouring can be dangerous, leading to dehydration. Your vet will prescribe a mineral and vitamin restorer to give as a drench.

If scouring is not the only symptom, check through the list above to see if you can identify the disease. It may well be due to worm infestation, in which case, it will clear up on its own once the worms are treated.

Scurf

Scurf is a common skin condition, occurring at the end of winter as the woolly undercoat starts to come away and the top coat begins to moult. Take one packet of flowers of sulphur and a small bottle of olive oil, and mix to a thin paste. Coat the affected areas with this mixture, leave it on for two to three days, then bath the goat. The loose scurf will come off, leaving the skin beautifully soft. This treatment is particularly useful for males that have dry, scurfy skin.

If a goat has scurf at any other time of year, it is probably due either to worm infestation or under-nourishment.

Treating for pink eye.

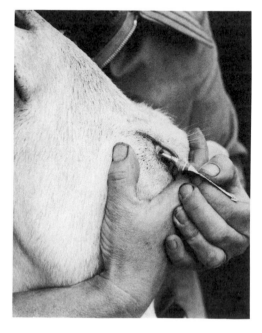

Dusting with louse powder for external parasites.

PARASITES

External

Lice can be destroyed and deterred with a cattle louse dressing. Ticks should be picked off carefully and destroyed (make sure that you have removed the head, too). Most other external parasites can be destroyed with a pesticide.

Internal

There is a rich variety of parasitic worms which can plague goats, although most of them can be kept at bay with a regular worming routine (*see* page 45). However, if you do not have access to a lot of ground for pasture, worms can become a constant problem and you will have to treat your animals more frequently.

The most common symptom of worm infestation is general lack of condition, and, in the case of the *Coccidia*, bloody scours. Occasionally, parasites reach the lungs causing parasitic bronchitis and professional advice should be sought in this case.

One parasite that is resistant to the normal worming treatments is the fluke, which is normally found only in marshy areas and which attacks the liver, causing wasting. A sample of the goat's droppings

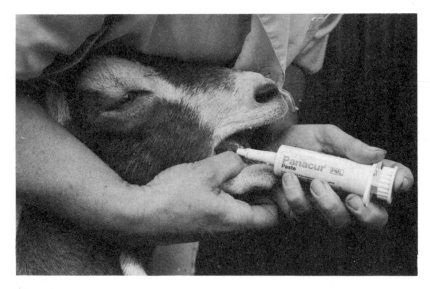

Worming.

analysed in a laboratory will confirm the diagnosis, and appropriate treatment can then be prescribed.

POISONING

Symptoms: collapse; colic; scouring.

Several common plants are poisonous to goats, but fortunately the majority of goats recognise which plants to avoid. If in doubt, check your pasture for any of these and dig them up (*see* opposite).

(a) Buckthorn
(b) Cowbane
(c) Deadly nightshade
(d) Dog's mercury
(e) Foxglove
(f) Greater celandine
(g) Hemlock
(h) Henbane
(i) Laburnum
(j) Laurel

(k) Privet
(l) Ragwort
(m) Rhododendron
(n) Rhubarb leaves
(o) Spindle
(p) Water dropwort
(q) Water hemlock
(r) White bryony
(s) Yew

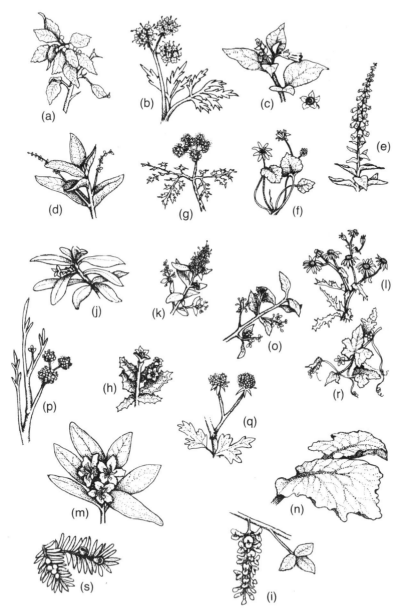

Plants poisonous to goats.

81

Other plants can be poisonous if large amounts are eaten over a period of time. This group includes bracken and green potatoes. Mouldy cereals and hay will not agree with your goats, but they are unlikely to eat them, anyway. Chemicals may also be the cause of poisoning.

If you think an animal has been poisoned, contact the vet and tell him the cause, if you can identify it, then drench the goat with 4fl oz (110ml) of liquid paraffin, followed by some black coffee or tea. Keep her warm until professional help arrives.

NOTIFIABLE DISEASES

Fortunately, none of the notifiable diseases are at all common, but you should at least know which infections have to be notified to the police. They are: foot and mouth, anthrax, rinderpest, cattle plague and rabies.

8 Showing

Showing your animals may not make your fortune – far from it! – but it is fun and gives you an opportunity to compare your goats with other people's and to 'chew the cud' with fellow goatkeepers. The fact that your animals may not be perfect specimens of their breed does not mean that they are not worth showing, although if they have some glaring faults in conformation or are below par, you would do better to leave them at home.

WHICH SHOW?

A large number of special goat shows are held all over Britain each year. Details of the major ones are printed in the British Goat Society's journal, but these are shows where only registered animals may be exhibited. You may feel that this is too 'big league' for you, in which case your local goat club will probably have its own show, and many small agricultural shows have a goat section.

CLASSES

These vary according to the size of show. One rule is that different sexes never compete, except perhaps for 'best in show'. Age also plays an important part. Female goats are divided into kids, goatlings that have not kidded, and goatlings or goats that have, and the male exhibits, into kids, and adults and bucklings. There will be separate classes for the different breeds, with the rest being grouped together under 'any other variety'.

The larger agricultural shows that have a goat section normally arrange milking competitions which start after the evening milking and go on for twenty-four hours. Indeed, many of the large shows are two-day affairs so if you are exhibiting only part of your flock, you will need to make arrangements for the care of those left behind.

Prize-giving at a small agricultural show.

SCHEDULES AND ENTRY FORMS

Assuming that you have decided to start by exhibiting your animals at your local affiliated goat club's show, you can obtain entry forms and the schedule, which lists all the classes and entry requirements, from your local club secretary. Make a special note of the closing date for entries, as this may well be eight weeks or so before the show date. Study the schedule carefully, to make sure that you are entering your animals in the class or classes for which they qualify, then give all the information requested on the entry form. If you are unable to fill in any section, explain why. Make sure that you post your entries in good time.

PREPARATION

If you have been following all the advice in this book, the amount of preparation required before showing your goats should be minimal, but there will still be some work to be done. You may not have trained your potential exhibit to walk to a lead properly or to stand still in a good position, so there will be a lot of hard grind before your first appearance in the show arena.

Training

All goats should be trained from an early age to wear a collar, since they will have to be tethered at various times (during milking, for example). You may even be used to taking them out for walks on leads, in which case the back of this task will be broken.

If you have never led a goat before, here are a few words of warning. There is no point in pulling on a goat as this will simply make the animal pull harder in the opposite direction. Also, in these circumstances you may be in for a shock, as the goat might fall, quivering, to the ground. Don't panic. All that has happened is that the strain on the collar has shut off the circulation in her neck. Leave her on the ground to recover and within no time she will be back on her feet and as good as new.

Remember that goats learn quickly if they are handled firmly – you have always to reinforce your position as leader of the flock – although there is no point in insisting that your animal carries on practising if she is tired or in a fractious mood. She may even prove temperamentally unsuited to exhibiting.

Just spend a few minutes a day with your animal, walking her on a loose lead, in a circle, then up and down, finishing up with her standing in a position so that all her good points are shown to advantage. The front legs should be straight and parallel, the hind feet planted very slightly behind the line of the rump and slightly further apart than the front feet, and the head held comfortably and not too high. A little physical encouragement may be needed to begin with, but your aim should be to resort to the minimum of control and direction – a touch on the lead, a tap on a leg should suffice by the end of training. Try to train her so that when you stand close to her with the lead in your hand and a finger against her neck she stands stock-still.

She also needs to become accustomed to being handled by a stranger – someone who will walk up to her and examine her mouth, pick up her feet and feel her udder. However calm she might be when you examine her, the experience of a stranger submitting her to these indignities may well make her lose her equanimity and with it any chance she has of winning an award. So, enlist the help of a few friends to put her through her paces.

The classic showing stance.

Grooming

If you have not been particular about daily grooming, now is the time to change your ways. A thorough brushing each day will help to produce an impressively glossy coat and, nearer the time, you can take other measures to improve your charges' appearance.

You will have to be extra fussy about keeping light-coloured goats and Swiss markings clean as the show date moves closer, which means mucking out the pens frequently, laying plenty of fresh straw each day, and removing any fresh stains from the hair as soon as you spot them.

A hoof-trimming session should be scheduled for a few days before the show, and you should check that the animal is standing squarely on all four feet after trimming. Clip the beard, if there is one, and the hair round the udder if necessary to give a neat outline.

On the day before the show, give your goats a good shampoo, taking care that they do not catch a chill. Wet the hair with warm water, rub in some horse or dog shampoo then rinse well. Shampoo again, working up a lather, then rinse off with copious amounts of warm water. Rub the animal down with a towel, taking care to dry a milker's udder thoroughly, smooth down the hair, then place a fresh towel on the goat's back and cover it with a goat rug. You should leave these in place until the animal is dry, then replace them with another rug

Bathing a kid in preparation for a show.

The bathed kid wrapped in a goat rug to keep her warm.

87

which should keep the goat clean until show time. The day before the show, check to make sure that there is no stained long hair. This will have to be trimmed off if it cannot be cleaned.

Feeding

During the lead-up to the show, take extra care with your exhibits' diet so that they are in peak condition on the day. Try to keep down the amount of concentrates that you give younger goats, as they should be trim rather than heavy-looking.

Packing

Packing up all your equipment for a weekend show can seem like preparing for an assault on Everest, and you would be wise to make a checklist. However, let's begin with your requirements for a less ambitious afternoon show.

Half-Day Show

Collars
Strong leads for tethering the goats to stakes
Light leads for showing
One bundle of hay per goat
One bundle of branches and greenstuff
Milking pails, if necessary
Grooming equipment
Tethering stakes
White overall
Warm jumper
Wellington boots
Stout mac
Show schedule, with instructions, passes and exhibit numbers, whichever are applicable

One-Day Show

As above, plus:
Extra food for goats, including concentrates
Buckets for watering and feeding
Bottles and teats for kids (if necessary) and means of heating their milk
Food for your lunch

Two-Day Show

As above, plus:
Even more food for goats
Even more food for you
Hay racks
Mineral lick brick
Bedding straw if not provided
Rugs
Cloth covering for the pens in case the night is cold
First aid box
Folding chair
Inflatable mattress or camp bed
Sleeping bag, blanket or quilt
Pillow
Change of clothing, plus extra socks

The packing list for the two-day event can be shortened if you plan to sleep in your van, but most exhibitors prefer to bed down alongside their animals' pen so that the goats feel more secure in their new surroundings.

GETTING THERE

Work out how long the journey should take, look at the schedule to see at what time you should arrive, then allow yourself at least an extra hour, depending on the distance. Make sure that the goats are fed a few hours before you set off and drive steadily and smoothly, at no more than 40mph (65km/h). If you have been allocated pens, settle the goats down as soon as you arrive and set up their hay and water supplies. If your goats are CAE free, your pens will be some way from the untested goats.

THE RING

Listen to the announcements carefully and have yourself and your goat ready to go into the ring as soon as the class is called. You will be told where to stand. If you are right at the bottom of a long line, don't make the poor animal stand in a perfect position right up until the judge reaches you. Do not let her frisk around, but do not be too particular until the judge is within sight.

Waiting for the judges.

MILKING COMPETITIONS

Milking competitions inevitably involve an overnight stay, since the goats are stripped out by the stewards in the early evening, after the owners have milked them, and the yield is measured over the following twenty-four hours. After stripping out, the milkers should be given their usual evening concentrates, since any variation in their diet will only upset them. Before owners and animals settle down for what will hopefully be a good night's sleep, the goats should be rugged, the hayracks filled, and the water buckets changed.

First thing in the morning, after a quick drink for the owners, breakfast for the goats and a freshen-up all round, the milkers go out into the ring to have their udders examined after which they have their first milking. Judging of all the classes in the show will take place, and sundry local and special cups and awards will be given.

The second milking is carried out in the late afternoon and the final results are usually worked out on the basis of the time that has elapsed since the last kidding. If any goat gives less than 5½lb (2.5kg) or if the percentage of butterfat in the milk is less than three at either of the two milkings, she is disqualified from the competition.

Show and Milk Recording Awards

The milking points may seem complicated enough to any new owner, but the British Goat Society's show and milk recording award system adds to the confusion.

Line-up of prize milkers.

The first awards to consider are the ★ and the Q★. A ★ is given to a goat that has gained at least 18 points in a milking competition, and whose milk had a butterfat content of not less than 3.25 per cent at either of the two milkings. A Q★ is given to a goat that has gained at least 20 points in a milking competition, and whose milk had a butter-fat content of not less than 4 per cent at either of the two milkings.

Then there are Challenge Certificates. There will be a Challenge Certificate for the best female goat plus Inspection-Production Challenge Certificates and Breed Challenge Certificates. A female champion must have won three Challenge Certificates under three different judges, and must have gained at least 18 milking points at the same shows. She must also hold three Inspection-Production Challenge Certificates with at least 18 milking competition points, in addition to having qualified for a Q★. To be entitled to be considered for a Challenge Certificate, a female goat must have gained at least 18 milking points and have qualified for a Q★. As for breed champions, a female goat must have won five Breed Challenge Certificates under three different judges with at least 16 milking competition points at the same shows and have been awarded a Q★ or a ★.

MALE GOAT SHOWS

Shows for male goats are held principally to enable breeders to evaluate the merit of stud males. As there are comparatively few males for show, other categories are often shown with them, particularly goatlings and kids.

The awards at a male show recognised by the British Goat Society follow similar lines to those outlined above. There are Challenge Certificates, and a champion male goat must have been awarded three of these under three different judges for best male goat over one year old. A male breed champion must have won four Breed Challenge Certificates under three different judges.

AFTER THE GOAT SHOW

You will probably not walk off with the red rosette at your first show, in which case you should learn from your mistakes. Try to work out what went wrong – pay attention to any comments by the judge, talk to your competitors and take a long look at those animals which gained a higher position than yours. Was it an irremediable flaw in your goat, or a shortcoming in presentation? Better luck next time!

9 Dairying

The lactation period of goats can be from nine to twenty-two months, depending on breed and age. Yield varies similarly from breed to breed and depending on how many times the goat has kidded – maximum yield usually occurs after the second and third kiddings. Interestingly, since goats have been bred specifically for increased milk production, some goats have been known to give milk even before they have their first kids.

MILKING EQUIPMENT

The obvious item is a milking pail, either enamelled or made of tin or stainless steel, with a lid and a 1 gallon (5 litre) capacity. Other essentials are a strip cup, a collecting churn and a strainer for straining the milk from the pail into the churn. You should also have a second

Milking equipment – strip cup and milking stool,
churn with strainer and stainless steel bucket.

strainer for straining the milk again when it is decanted into manageable containers for cold storage. You will need access to a sink with running water for cooling the churn, a rack for draining the equipment after washing, and a table. You should obtain a hanging scale to keep track of milk yields.

Any other equipment depends to some extent on the scale of your operation and on whether you intend to sell some of the milk. For commercial milking you would need a large cupboard for storing bottles or cartons, a big refrigerator, and some kind of bulk milk-cooling system.

WHEN TO MILK

You can start to milk your goat from the fourth day after she has had her kids and you *must* milk her at regular intervals, twice a day. A simple calculation will tell you that this means every twelve hours, or as near as you can make it. One of the reasons for this is that regular milking at evenly spaced intervals ensures a consistent butterfat content, but your goats will also be more comfortable, and therefore happier, if a routine is adhered to. Some goatkeepers prefer milking at ten- and fourteen-hour intervals for convenience, bearing in mind the short winter days, and this is perfectly acceptable, although you will have to put up with variations in butterfat content.

BEFORE MILKING

There are a couple of very important 'don'ts' to observe before milking. First, never feed your animals anything with a strong flavour – kale or turnips, for example – within four hours of milking, as the flavour will be passed into the milk. Second, if you are milking in the goathouse, do not sweep up beforehand, as this will create clouds of dust particles which will take a long time to settle and which will inevitably contaminate the milk.

APPAREL

The most important thing to remember is hygiene. As mentioned earlier, goats' milk does not need pasteurising, so you must be very particular about hygiene at all times. Routine is just as important in the preparation stage as it is during milking. It might be a good idea to have a special overall that you wear for milking, but that doesn't mean that you can simply sling it on after mucking out and all will be well. Your clothes must be clean and your hands and arms should be scrubbed down – rather like a surgeon preparing for an operation. You may not worry about your family swallowing down the odd germ or two, but if you are contemplating selling your milk or offering it to friends, it pays to be scrupulous about hygiene.

'LET-DOWN'

'Let-down' is an expression which will be very familiar to nursing mothers, but perhaps less so to other people. It is a physical reaction to a particular trigger which releases the sphincters and valves which hold the milk in the udder. The trigger mechanism will vary, depending on your routine. It may be the clanking of the milk-pail, or a certain tune you whistle as you are getting ready to milk (in which case, remember to whistle the tune every time!), or, most commonly, it is the feeling she has when her udder is wiped and massaged gently. Milking your goats in the same order every time can also encourage the let-down reaction at the appropriate moment.

HOW TO MILK

We have all seen those films where a complete novice makes comic, and fruitless, attempts at milking an animal but, in reality, the knack is quickly acquired.

Before tackling your animal, you must wipe her udder with a clean cloth or udder wipes. This, followed by a brief, gentle massage, will encourage let-down. Wrap your fingers around the teat, with your thumb uppermost, making sure that your fingertips do not touch the teat. Press gently upwards into the udder to release some milk into the teat, then smoothly and firmly exert pressure with each finger in turn, starting from the top, in a sort of wave motion, forcing the milk down the teat and into the pail. Relax your grip and the upwards pressure on the udder, then repeat the process using your other hand on the other teat.

The first four jets should be caught in a strip cup, as they contain any germs or bacteria which were lodged in the teats. Before throwing this milk away, check for flakiness or any stringy quality which could be the symptom of mastitis, or some other disease. As the udder gradually empties, you will have to increase the upward pressure of your

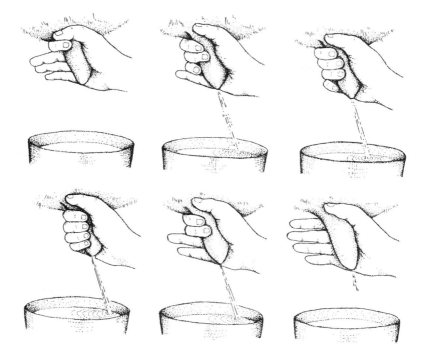

*Milking – you should exert pressure on the teat in a
wave motion.*

hands, and eventually the milk flow will stop. This does not mean that
the udder is empty. Massage the udder gently from the top and the
back in a downwards direction towards the teats, and start milking
again. Repeat this sequence until you have extracted every last drop –
the last milk is the richest, and a goat that is consistently undermilked
will respond by producing less.

Speed is of the essence when you are milking a goat. If there is too
long a pause in the process, the goat's sphincters will tighten and you
will get no more from her this time. One of the most important points
for the novice to remember is that a goat is not a cow, and her teats
should not be treated like bell-ropes. If you pull downwards during
milking, you will eventually distort the udder.

AFTER MILKING

As soon as you have finished milking the goat, weigh the results on the hanging scale, make a note of them, and then pour the milk through a strainer into the churn. This should be standing in cold running water to cool the milk as quickly as possible, since goat's milk develops an unpleasant flavour if it is not cooled quickly, and warm milk is also an ideal breeding ground for bacteria.

By the time you have finished milking and cleaning up, all the milk should be at tap temperature, and it should then be strained into bottles or other containers and kept in the fridge. If you find yourself with a slight surplus in the late summer, the milk will survive perfectly happily in the freezer for up to three months.

CLEANING UP

Once all the goats have been milked, clean every piece of milking equipment. Begin by rinsing the various items in cold water, then add some dairy detergent to warm water, and fill, cover and wash the equipment. Rinse with plenty of water, then scald with boiling water and leave upside-down on a rack to dry. Check very carefully that you have washed away every last trace of milk. To make doubly sure that no bacteria build up and contaminate the milk, you should sterilise the equipment once a day, either by soaking it in a sterilising solution, steam-treating it, or by heating it in an oven.

MILK YIELDS

The purpose of recording your goats' yields is not so that you can bask in a warm glow as you see the quantities rise, but rather so that you can assess the effect of your management on the animals. All goats tend to give less milk during the winter, but injudicious exposure to the elements at other times of the year may affect milk production, as may worms, diseases, variations in feed, and psychological stress. Yield records are a valuable barometer for monitoring the condition of your animals.

MILK RECORDING

There is a scheme of milk recording which is run under the auspices of the Milk Marketing Board. The amount of milk from each goat in the scheme is weighed a set number of times during each lactation by an official from the MMB, and the butterfat and protein content is checked. The goat owner must weigh and note down the milk from each goat each day. At the end of each lactation the goat may qualify for certain awards.

A similar scheme is run by goat clubs affiliated to the British Goat Society. This is a cheaper scheme, and is called club milk recording. Details of these schemes and the awards may be obtained from the British Goat Society.

DAIRY PRODUCE

If you have two milkers producing up to 8 pints (5 litres) of milk a day each, you will probably have a surplus, so it is worth considering some of the products you can make with that excess milk. Those that immediately spring to mind are butter, cream, yoghurt and cheese.

Cream

Cream does not separate off goats' milk as readily as it does off cows' milk simply because the fat globules are smaller, and the milk is therefore, to some extent, self-homogenising. However, if you follow the directions given in the butter section below, you should be able to obtain a satisfactory quantity. The end result is more digestible than cows' milk cream and whips up to a greater bulk, although it is, to some, an unappetising colour – pure white.

The only problem if you take off the cream is that you will be left with a large amount of skimmed milk. However, this is ideal for young animals – and in cooking, if you are worried about your calorie intake!

Butter

In many people's opinion, making butter from goats' milk is not a worthwhile undertaking. The colour of goats' milk butter is unappetisingly like that of lard (you can add a little annatto colouring to solve this problem), but there are those who swear that its taste is vastly superior to cows' milk butter.

The most difficult part is separating the cream off. If you have a separator this is an easy task; if not, the method is as follows. Pour some milk into a wide, shallow pan, so that there is a large surface area exposed to the air, then leave it to stand in a cool, dust-free place for twenty-four hours. You should then be able to skim off the cream and store it in the fridge. Repeat this process with a fresh batch of milk every day for a week, and you should have enough to make a worthwhile quantity. Leave the last skimming of cream at room temperature for a day to ripen, then put the week's collection into a mixing bowl and whisk with a mixer until the butter forms. You can even put the cream in a large jar and shake it for a quarter of an hour or so, but this technique is more tiring.

When the butter has formed, be sure to drain off the buttermilk (it can be used in cooking), otherwise the butter will turn rancid. Work the butter with a wooden spoon, pressing it against the sides of the mixing bowl and pouring off any buttermilk that is exuded, or squeeze out any remaining buttermilk through a muslin cloth. Rinse the butter several times in cold water to remove any traces of milk until the rinsing water remains clear, and pat off the moisture with a cloth. If you prefer your butter salted, add some salt (1 tsp per lb [450g]) and work it in before shaping.

Yoghurt

As those who already make home-made yoghurt (from cows' milk) will know, there is no comparison between this and the shop-bought variety. You can use exactly the same method for goats' milk yoghurt.

You will need a starter – either a commercial culture at 4oz to 1 gallon (100g to 5 litres) of milk, or one small carton of unpasteurised shop-bought yoghurt for every pint. (For later batches you can use yoghurt from the preceding production as a starter, although its potency will gradually weaken and you will need to use commercial yoghurt or culture again.)

Heat the milk to 150°F (65°C), then cool it rapidly to 115°F (46°C). Pour three-quarters of the milk into an insulated container (a thermos for small quantities, and an improvised insulated container for larger quantities), and mix the culture or yoghurt starter into the remaining milk. Add this to the rest of the milk, mix thoroughly, then leave to stand for four to six hours, by which time it should be set. Open the container to the air and allow the yoghurt to cool and stand for ten to twelve hours before use.

Cheese

Cheese is probably the best known of all the goats' milk by-products, and the only equipment you will need beyond what is already in your kitchen is rennet, a cheese thermometer and some muslin.

Cottage Cheese

This is the easiest cheese to produce. Simply heat up some milk, say 1 gallon (5 litres) to 90°F (32°C). Dissolve some rennet in cold water (follow the directions on the pack), turn off the heat, and stir the rennet solution into the milk. Leave the mixture to stand in a warm place for about an hour until a curd forms. Then cut the curd into small cubes and stir very gently. Heat the curds and whey very slowly until they reach 110°F (43°C) and, when the cheese is as firm as you want it, strain the whey off through a sieve lined with muslin. This quantity of milk should produce about 2lb (1kg) of cheese.

If you don't want to use rennet, you can buy some starter culture from a dairy laboratory, or even crumble a piece of unprocessed cheese into the milk, in which case you will have to leave it to stand in the warm for a day.

Hard Cheese

Using more or less the same method, you can produce a very palatable hard cheese. The quantity of cheese produced will, however, be less so you may want to start off with twice the amount of milk. Heat the milk, add the rennet and leave to stand and cut the resulting curd into small cubes, as before.

Now heat the contents of the pan *very* slowly indeed, until they reach 100°F (38°C) and keep them at that temperature until the consistency is as firm as you want it, stirring very gently from time to time. Be warned, this can be a lengthy process. Then drape a large square of muslin over the mouth of a pan, transfer the curds and whey to this pan, lift them up to the muslin and leave to drain. When all the whey has dripped out crumble the curd and mix in some salt.

It is at this stage that you need a cheese press. If you do not have one, a tin can with both ends removed will do or, for a large quantity, you can use a loose-bottomed cake tin.

Assuming that you are using a tin, put the curd in the tin, lined with muslin, then cover with muslin and a wooden disc (called a follower) the size of the opening. Pile on 20lb (10kg) weight (you might need a lever system to exert this kind of pressure on a relatively small area)

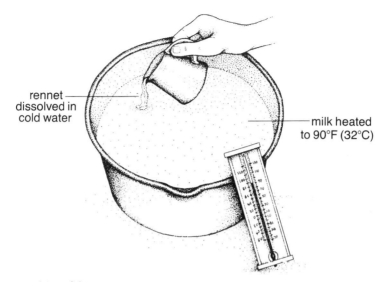

rennet
dissolved in
cold water

milk heated
to 90°F (32°C)

*Cheese making. (a) Stir dissolved rennet into milk
heated to 90°F (32°C).*

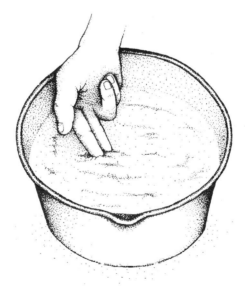

(b) Leave to stand until curds form.

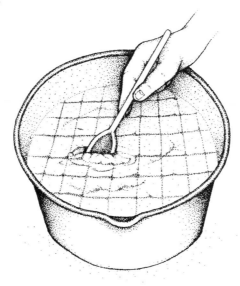

(c) Cut curd into small cubes and stir gently.

(d) Heat slowly until contents reach 100°F (38°C), and keep at this temperature until consistency is suitably firm.

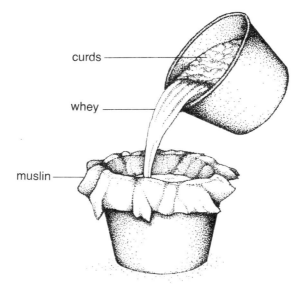

curds

whey

muslin

(e) Transfer curds and whey to muslin covered pan.

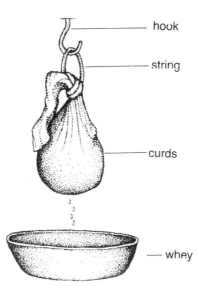

hook

string

curds

whey

(f) Lift up muslin and leave to drain.

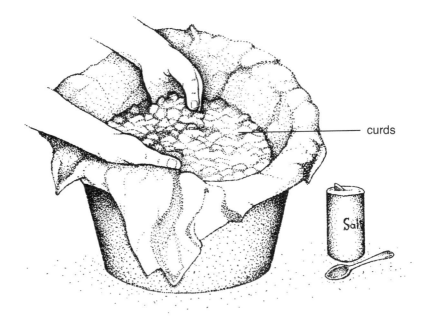

curds

*(g) When all the whey has dripped out, crumble the
curd and mix in some salt.*

and leave for twenty-four hours. At the end of this time, remove the
cheese and steep it in water for fifteen seconds at 150°F (65°C) to
encourage the rind to form. Wrap the cheese up in some fresh muslin,
put it back in the tin the other way up, and apply twice the pressure.
The amount of pressure is by no means critical. If you increase the
pressure you get a harder cheese, and with light pressure the result
will be softer.

Remove the cheese after twenty-four hours, wipe it with a clean
cloth and seal any cracks by applying some warm water and smooth-
ing the surface with your finger. Wrap the cheese up in some fresh
muslin, and dip the cheese in melted lard to keep in the moisture. The
cheese should now be stored in a cool place (not a refrigerator) for
anything from one to two months, and should be turned several times
daily. You will find a mould appearing as time passes – simply clean it
off with a brush or salt water.

Glossary

Buck A male goat.

Buckling A male goat between one and two years old.

Conformation The physical characteristics of a breed to which an animal of that breed ought to conform.

Cudding The action of masticating the regurgitated contents of the rumen prior to final digestion in the abomasum.

Doe A female goat.

Drenching Giving a goat a liquid medicinal treatment in a bottle.

Goatling A goat that is in its second year.

Heck A wooden structure part of which contains fodder, while the other part serves as a restraint for the animal during feeding.

Hermaphrodite A goat that has neither the complete reproductive organs of the female, or of the male – more commonly seen in hornless goats.

Kid A goat that is less than one year old.

Registered stock Goats which are registered with the BGS.

Scouring Diarrhoea.

Stripping Taking all the milk out of a goat's udder.

Swiss markings Facial stripes from the top of the head down to the muzzle, around and inside the ears, down the legs and on the rump. These are usually white and appear on certain breeds.

Further Reading

Belanger, J., *Raising Milk Goats the Modern Way* (Garden Way Publishing, Vermont, US, 1975).

The British Goat Society, *The Herd Book*
The Monthly Journal
The Stud Goat Register
The Year Book

Halliday J. and Halliday J., *Practical Goat Keeping* (Ward Lock, 1987).

Holmes Pegler, H.S., *The Book of the Goat* (The Bazaar Exchange & Mart Ltd, o/p).

Jeffery, H.E., *Goats* (Cassell & Co. Ltd, 1970).

Mackenzie, D., *Goat Husbandry* (Faber and Faber Ltd, 1980).

Neal, Jenny, *Goatkeeping for Profit* (David & Charles, 1988).

Rogers, F., *Goats, Their Care and Breeding* (KR Books Ltd, 1979).

Salmon, J., *The Goatkeeper's Guide* (David & Charles, 1981).

Shields, J., *Exhibition and Practical Goatkeeping* (Saiga, 1983).

Thear, K., *Commercial Goat Farming* (Broad Leys Pub. Co., 1985).

Goats and Goatkeeping (Merehurst, 1988).

Useful Addresses

ORGANISATIONS

British Goat Society
34–36 Fore Street
Bovey Tracey
Newton Abbot
Devon

Goat Producers Association
Mrs J. Barley, Secretary
c/o A.G.R.I.
Church Lane
Shinfield
Reading
Berkshire

Goat Veterinary Society
J. Matthews, Secretary
The Limes
Chalk Street
Rettendon Common
Chelmsford
Essex

MAGAZINES

The B.G.S. issues a very useful monthly journal called *Goats Today*.

Smallholder
Liz Wright, Editor
Tel. 0354 741182

A very useful monthly magazine dealing with all types of livestock as well as goats. Available from booksellers.

Useful Addresses

Country Garden
Buriton House
Station Road
Newport
Saffron Walden
Essex

Six issues per year, by subscription only.

EQUIPMENT

W. N. Boddington & Co. Ltd
Horsmonden
Kent

Suppliers of cheese moulds.

Christian Hansen Laboratories Ltd
476 Basingstoke Road
Reading

Suppliers of rennet and cheese cultures.

Index

Index

Other titles available from The Crowood Press

A Guide to Management Series

Ducks and Geese   Tom Bartlett

Growing Herbs   Rosemary Titterington

Organic Farming and Growing   Francis Blake

Pigs   Neville Beynon

Poultry   Carol Twinch

Profitable Free Range Egg Production   Mick Dennett

Sheep   Edward Hart